Robert Lenzner is able to unravel the accumulation of this enormous fortune after years of research. He is well qualified to do so, being an American correspondent of the *Economist* as well as New York bureau chief of the *Boston Globe*. Among the masses of documents he uncovered, some revealed for the first time, are damning security reports on Getty by the FBI, and the twenty-one codicils he made to his will in the last eighteen years of his life. Among the people who have talked frankly to him are Getty's family, employees, social acquaintances, friends and members of his harem. The enigma may never be solved, but this book contains all the clues.

ROBERT LENZNER

Getty

The Richest Man in the World

GRAFTON BOOKS

A Division of the Collins Publishing Group

LONDON GLASGOW
TORONTO SYDNEY AUCKLAND

Grafton Books
A Division of the Collins Publishing Group
8 Grafton Street, London W1X 3LA

Published by Grafton Books 1987

First published in Great Britain by
Century Hutchinson Ltd 1985

Copyright © Robert Lenzner 1985

ISBN 0-586-07147-4

Printed and bound in Great Britain by
Collins, Glasgow

Set in Times

To Nina Rachel Lenzner,
the finest

Contents

Illustrations

Paul Jr with Talitha Pol
Teddy, Getty's fifth wife, and their son Timmy
Getty with Rosabella Burch
The J. Paul Getty Museum in Malibu, California
Getty at Sutton Place with Marianne von Alvensleben
and Norris Bramlett
Getty with his lion cub, Nero
Claire Getty with her grandfather at Sutton Place in 1975
Talitha with Tara in 1971
Gordon and Ann Getty at the Orchid Ball in Los Angeles
in 1984
Paul III with his wife Martine and his son Paul Balthazar
and Martine's daughter Anna
Gail Getty, Paul Jr's first wife, with their children, Mark,
Ariadne and Aileen

Acknowledgements

This book could not have been completed without the aid of many remarkable people. First and foremost, the editors of the *Boston Globe*, and in particular Tom Winship, who allowed me to take many months away from the grind of daily journalism; Harold Harris, who skilfully reshaped the manuscript; Roddy Bloomfield, who lit the fire to get the book out; Bill Phillips, who helped devise the theme for the book; Sue Hogg, who so ably made sense out of the manuscript; John Ware, who insisted that the book outline be perfect; Vince McCullough, who was a crucial aid in rewriting the book; and Susan Baum, who organized me and the material so that every deadline could be met. Also, I want to thank all my researchers in New York, Los Angeles, Mexico, Rome and London. Many wise and warm friends gave me their advice, assistance and, most importantly, encouragement: Amram Ducovny, Leon Levy, Martin Lipton, Peter Martin, Martin Sosnoff and Michael Thomas.

I met the most extraordinary group of people who helped give me access to Getty's life. First, Alexander Papamarkou, who kindly led me to Ann Getty, who in turn made it possible for me to meet her husband, Gordon. Other members of the family who contributed their insight include, most prominently, J. Paul Getty Jr and his son Mark, and Mark's mother, Gail Getty. A close Getty family friend, Bill Newsom, and the Getty family lawyers, Moses Lasky and Edward Stadum, were immensely helpful and thoughtful throughout. So was J. Paul Getty Jr's lawyer in London, Vanni Treves. This book, however, was not authorized by any Getty or approved by any Getty. While I respect the opinion of the Getty family, I have written the story as close to the truth as was possible to discover, given the incredible complexity of J. Paul Getty's life.

Some Getty friends and acquaintances stand out for their co-operation and illuminating insights into the fascinating and contradictory character of J. Paul Getty. I would like to thank in particular Jeannette Constable-Maxwell, Claus von Bulow, Paul Louis Weiller, the Duchess of Argyll, the Duke and Duchess of Bedford, Mary Teissier, John Brealey, David Staples, Everette Skarda, Giorgio Schanzer, Frederico Zeri, Stephen Garrett, Marianne von Alvensleben, Hilde Kruger, Harold Berg, Norris Bramlett, Anna Hladka, Teddy Gaston, Ann Light, Stewart Evey and Barbara Wallace. Getting inside Getty would have been impossible without their pungent, intimate knowledge of the man.

Photographic acknowledgements

For permission to reproduce photographs included in this book the publishers would like to thank Associated Press, the BBC Hulton Picture Library, Camera Press, Desmond O'Neill Features, Popperfoto, Rex Features, Texaco and UPI/Bettermann Newsphotos.

The plutocracy, in a democratic state, tends to take the place of the missing aristocracy, and even to be mistaken for it. It is, of course, something quite different. It lacks all the essential character of a true aristocracy: a clean tradition, culture, public spirit, honesty, courage – above all, courage. It stands under no bond of obligation to the state; it has no public duty; it is transient and lacks a goal . . . Its main character is its incurable timorousness; it is forever grasping at straws held out by demagogues . . . Its dreams are of banshees, hobgoblins, bugaboos.

H. L. Mencken

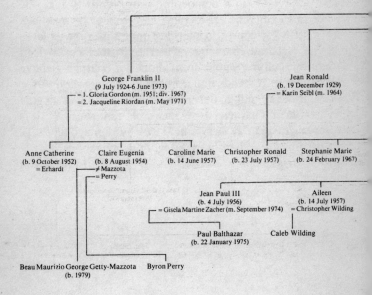

George Franklin II
(9 July 1924-6 June 1973)
= 1. Gloria Gordon (m. 1951; div. 1967)
= 2. Jacqueline Riordan (m. May 1971)

Jean Ronald
(b. 19 December 1929)
= Karin Seibl (m. 1964)

Anne Catherine
(b. 9 October 1952)
= Erhardt

Claire Eugenia
(b. 8 August 1954)
≠ Mazzota
= Perry

Caroline Marie
(b. 14 June 1957)

Christopher Ronald
(b. 23 July 1957)

Stephanie Marie
(b. 24 February 1967)

Jean Paul III
(b. 4 July 1956)
= Gisela Martine Zacher (m. September 1974)

Aileen
(b. 14 July 1957)
= Christopher Wilding

Paul Balthazar
(b. 22 January 1975)

Caleb Wilding

Beau Maurizio George Getty-Mazzota
(b. 1979)

Byron Perry

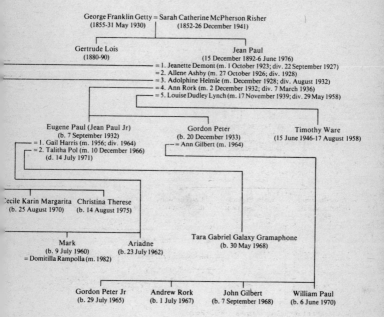

George Franklin Getty = Sarah Catherine McPherson Risher
(1855-31 May 1930)   (1852-26 December 1941)

Gertrude Lois   Jean Paul
(1880-90)   (15 December 1892-6 June 1976)
            = 1. Jeanette Demont (m. 1 October 1923; div. 22 September 1927)
            = 2. Allene Ashby (m. 27 October 1926; div. 1928)
            = 3. Adolphine Helmle (m. December 1928; div. August 1932)
            = 4. Ann Rork (m. 2 December 1932; div. 7 March 1936)
            = 5. Louise Dudley Lynch (m. 17 November 1939; div. 29 May 1958)

Eugene Paul (Jean Paul Jr)   Gordon Peter   Timothy Ware
(b. 7 September 1932)   (b. 20 December 1933)   (15 June 1946-17 August 1958)
  = 1. Gail Harris (m. 1956; div. 1964)   = Ann Gilbert (m. 1964)
  = 2. Talitha Pol (m. 10 December 1966)
       (d. 14 July 1971)

Cecile Karin Margarita   Christina Therese
(b. 25 August 1970)   (b. 14 August 1975)

                                          Tara Gabriel Galaxy Gramaphone
                                                  (b. 30 May 1968)
        Mark   Ariadne
   (b. 9 July 1960)   (b. 23 July 1962)
 = Domitilla Rampolla (m. 1982)

    Gordon Peter Jr   Andrew Rork   John Gilbert   William Paul
   (b. 29 July 1965)   (b. 1 July 1967)   (b. 7 September 1968)   (b. 6 June 1970)

Preface

'Why are they sending captains to see generals?' J. Paul Getty asked me in a deep, intimidating tone that August day of 1965, when I phoned him from a call box in Guildford, England, in an attempt to gain an interview. I was an investment banker then, eager to persuade him to allow my firm, Goldman, Sachs & Co., to represent him in selling oil properties worth $300 million. Nevertheless he invited me to his home.

What I remember of Sutton Place was the sharp contrast between the earthern floor of the gatehouse and the spectacle of tapestries, portraits and furniture in the never ending Long Hall; between the forbidding pens of the Alsatian guard dogs and the tulip-shaped yew trees of the formal garden. My guide was a pretty auburn-haired assistant wearing too much make-up.

After the tour I was led into a small study, and presently a stooped but impeccably dressed Getty entered. There was little small talk while for half an hour he grilled me about my proposition, his cold blue eyes showing such total concentration that he made me ill at ease. He listened with a bone-dry seriousness, offering little of an opening. He was most perturbed that the US Justice Department had blocked his attempt to sell oil properties to Standard Oil of New Jersey – once an archenemy of him and his father. He thought that British law would have been fairer to him. Oddly, he was annoyed that Standard Oil stock sold at a higher dollar price per share than Getty Oil shares.

For me, at the time, he was a legendary figure, living

an ocean and a continent away from his corporate headquarters. There was no trace of his birthplace, Minneapolis, or southern California, his longtime home. Because he was so different from the usual chairman of the board, Getty was like a fictional character, but not a shady arriviste like Jay Gatsby, nor a flamboyant spendthrift like Citizen Kane, nor even a sweet-tempered billionaire art collector like Adam Verver.

When he died in June 1976, I was in London on assignment for my newspaper, the *Boston Globe*, and, along with everyone else, enjoyed reading about the numerous women to whom he left money. Never before had a wealthy man publicly left so many different women money in his will.

I had read his autobiographies and intuitively felt they did not tell the whole story or, for that matter, the truth. I decided to try to find the real Getty, to explain how he made his unlimited wealth and what this effort did to him and those around him.

In searching for the real Getty I employed one of his favourite homilies: 'No man's opinions are better than his information.' To that end I interviewed his colleagues, friends, lovers, ex-wives, family, competitors and employees. I searched the courts for records of his personal and professional lives. Government files, long hidden away, revealed nooks and crannies of his life that he had never made public.

What struck me more than anything else was the enigmatic quality in Getty, the puzzling contradictions in every aspect of his life, whether it was oil, art, ladies or family. What made him unique were the many Getty personas – Getty, the hardbitten oilman; Getty, the enthralling lover; Getty, the absent parent; Getty, the miserly art collector; Getty, the philandering husband; Getty, the bohemian expatriate; Getty, the social snob;

Getty, the physical coward; Getty, the naïve Midwestern hick; Getty, the puritan; and Getty the writer, who wanted to teach everyone how to be rich. That he was all these things makes him a fascinating figure, highly unusual for a businessman. What ensures his place in the history books is the creation of America's largest single family fortune which made J. Paul Getty the Richest Man in the World.

1
Lying in State

I never realized I'd be an orphan one day.

J. Paul Getty, 1975

The coffin of Jean Paul Getty, illuminated by huge flickering beeswax candles, rested in the middle of the Great Hall of Sutton Place, flanked by silent security men who stood as if guarding not a corpse but the man's entire fortune.

In his obsession with personal safety, J. Paul Getty, who had spent the last twenty-five years of his life in exile from his native America, had turned his seventy-two-roomed house into a private fortress. A dozen Alsatian dogs, trained to tear intruders to pieces, roamed the grounds. Bars obscured the view from each of the mansion's five hundred windows and, at night, armed guards had been stationed outside the closed doors of the master bedroom. But now they had a lonely vigil.

On that sweltering day in June 1976 no convoys of Rolls-Royces or Bentleys carrying dukes and duchesses, oil barons, press lords, art dealers, politicians and film stars came to pay their last respects to the man whose hospitality in his lifetime they had freely accepted. It was 'the saddest sight' Getty's old friend, Commandant Paul Louis Weiller, 'had ever seen. Paul was all alone, without friends or family, only a few servants, no one else. I could not believe it.'

A pair of his adoring lady friends made their way to the sombre Great Hall to view the body. Rosabella Burch, a Nicaraguan beauty who had been supported by

Getty for years and who wanted to marry him, tearfully placed a pin in the form of a crucifix on his body, before being ushered out of the door. Next, Anna Hladka, from Czechoslovakia, an art researcher and one of Getty's lovers in later life, stood looking at the man who was 'dearest to me in the world, who never used perfume, but was clean and sweet, with the body of a young child.'

Getty would have been delighted, however, at the turn-out of titled and notable acquaintances who attended his memorial service on 21 June 1976 at the American Church of St Mark in North Audley Street, London. Lord and Lady Southborough, his neighbours in Surrey, C.Y. Tung, the Chinese shipping tycoon, and Joseph Floyd, the distinguished chairman of Christie's, joined a sprinkling of earls, countesses, lords, ladies and family to honour his memory.

Of his three remaining sons, only Jean Paul Getty Jr, estranged from his father for the last five years, was there, and he arrived late, wearing dark sunglasses and white sneakers, with Bianca Jagger on his arm. It had been hard for Paul Jr to leave his place of refuge at Cheyne Walk, Chelsea. He was 'deeply disturbed and hurt' by his father. But he came because 'I loved him, I didn't hate him.'

In the congregation also was Paul Jr's ex-wife Gail and their four children – J. Paul III, who had lost an ear during his notorious kidnapping three years earlier, his brother Mark and his two sisters, Aileen and Ariadne. Although Gail was bitter at the old man's delay in paying the ransom to free her eldest son after six months in captivity, she recalls him as being a 'sweet man'. Over the years, she discovered that Getty 'had a different persona around different people. He could be very, very charming when he felt comfortable with people, but the persona around business people was a tough one. I'm

Paul Getty, and I run this show.' To Gail, Getty the art collector and Getty the lover were different from Getty the father and Getty the oil tycoon.

The funeral oration was delivered by Getty's friend, the spare, urbane Duke of Bedford, a symbol of the British nobility over which Getty fawned. Getty could not have planned a better eulogy himself. He had no wish to be remembered as he really was. He was a mythmaker. He wanted to be the richest man in the world but liked to pretend that he 'never felt rich'. At the end of his days the thin, stooped oilman bore a curious resemblance to John D. Rockefeller, the robber baron who founded Standard Oil Trust. Getty's lips, thin as a Rockefeller dime, were the expression of his parsimony; his long, imperial nose and cold, expressionless blue eyes betokened his power and aloof reserve. Yet, Bedford told Getty's admirers, 'He was not a mean man, as is often written about him. On the contrary, he was the kindest and most generous of friends, as I am sure many of you will agree. He just had an allergy to money badly spent or badly invested or lying idle. I imagine one does not become a billionaire without this discipline.'

There must have been many wry smiles on the faces of those in St Mark's who heard the Duke's address and who had first-hand knowledge of Getty's 'allergy' when it came to spending money. One of them was Lady Diana Cooper, whom Getty had often squired to social and cultural functions. On one occasion, when she complained of rheumatism, Getty told her he had a first-class remedy and promised to send her some. He was as good as his word. Several days later she received a used tube of Bengue's Balsam, empty, shrivelled, with the neck of the tube bent over to squeeze out every last ooze. The Duke of Bedford himself had experienced Getty's 'allergy'. When dining at Sutton Place he had raised his eyebrows

after dinner when the dining-room doors were ceremoni-
ously closed and locked, as if the host expected some
member of the invited party to make off with the silver.
'When I think of Paul, I think of money,' says the Duke.

During the last eighteen years of his life Getty had
tantalized his lady friends, his family and personal staff
with hints or even outright promises of what they could
expect from his will. It had been changed no less than
twenty-one times, with legacies being reduced or even
cancelled if the beneficiaries had somehow offended him.
It was his personal method of reward and punishment.

Throughout his life Getty maintained relationships with
many women. Some were lovers; others were just friends.
He gave money in varying amounts to both lovers and
friends alike. Only one of his lady friends who attended
the service in St Mark's came close to being made a
millionairess on his death. Penelope Kitson was a tall,
cool-headed British interior designer who had worked for
Getty on the refurbishment of Sutton Place and was given
a cottage on the estate. He left her 5000 shares in Getty
Oil, worth $825,000, plus a monthly allowance for life of
$1167. Penelope was the one woman who had managed
to remain emotionally independent from him and, as a
result, she thought, he was fascinated 'by someone he
could not control'.

Robina Lund, Getty's public relations consultant and a
director of Sutton Place Properties, had once shared top
ranking as a potential beneficiary. In 1967 he had planned
to leave her 6250 shares in Getty Oil which would have
been worth $1 million when he died, together with a
monthly cheque of $750. In the end there were no shares
for her, only a paltry $209 a month. To Robina Lund,
Getty was 'a man of complete opposites; for every quality
he had the balancing fault'.

It was true that Getty's character was a mass of contradictions and inconsistencies. He could throw a $28,000 party, and still put a pay telephone in his house. He could be an enormously satisfying lover, and yet dismiss a person from his life without a shrug. He had the most perfect and impeccable manners, but they were superficial, concealing the rough and primitive human being who lived beneath the surface. He would insist on not breaking the law in business, but seemed to be totally unaware of the codes that rule our personal lives. He could be more daring than most other entrepreneurs, but was obsessed by fears and was unable to return to his native land because he could not face flying.

Perhaps Getty was incapable of a lasting, intimate relationship, but he was enormously shrewd about the details of amassing wealth and power. For some forty-six years, longer than any other American businessman in the twentieth century, he never relinquished the presidency of Getty Oil and, from thousands of miles away, made his will known in the oilfields and office towers of his empire. Jeannette Constable-Maxwell, who mourned the loss of a long-standing friend, recalls, 'There was no man less idle. He never coasted; he always operated in low gear, but he was an inexorable mover. There was not a hill he couldn't climb.' She would miss his sudden appearances in his chauffeur-driven Cadillac at her home on the Chelsea Embankment to tell her the latest funny stories he had heard.

Getty always knew that his fortune was on everyone's mind. He must have foreseen, with sardonic humour, the widespread disappointment when it was revealed that he had left almost his entire fortune to a museum in California which bore his name but which he had never visited. This copy of a first-century Roman villa expressed the grandeur of Getty's imperial fantasy. It was his greatest coup. It was also a symbol of the mass of contradictions in Getty. In his lifetime he could not bring himself to pay high prices for masterpieces, but bought only bargains, many of them second- or third-rate. But in death, he left unlimited funds to a museum already filled with his personal art collection. 'He wanted to make sure his name would be perpetuated as long as there was civiliz-ation,' was the view expressed by Norris Bramlett, Getty's faithful Man Friday and personal assistant.

Throughout his life, Getty had put a financial value on everything, even love and companionship, and many of those who attended that memorial service at St Mark's must have wondered about the calculations that lay behind some of his bequests. Even Penelope Kitson, the most favoured woman friend, believed, in her own words, that 'he was frightfully wrong in the will. He hurt a great many little people.'

He even succeeded in depriving the British tax man of death duties. Sutton Place itself belonged to a subsidiary of his American company and most of his personal art collection had already been shipped to California. As for his fortune, the British Labour Government of Harold Wilson had passed a law in 1975 aimed at the estates of wealthy foreigners living in England. It made them liable for death duties if they had been resident in the country for seventeen of their last twenty years. Getty, who had lived in self-imposed exile like some dethroned king, managed things with immaculate timing. One year after the bill became law, he died at the end of his sixteenth year at Sutton Place.

And so at last the billionaire returned home to Califor-nia in the cargo hold of a TWA jumbo jet on a scheduled flight from London's Heathrow Airport. He left behind one other man who must have suffered a slight financial disappointment. The payment which had been promised to the minister at St Mark's Church was not forthcoming. Getty got his memorial service for free.

2

The First Two Decades

After I had seen the toy department I went to the book
department where I got two books, *Bound to Rise* and *Risen
from the Ranks* by Alger.

J. Paul Getty, 26 October 1904

That J. Paul Getty was born at all was a minor miracle.
Two years earlier, in 1890, a sister, Gertrude Lois, whom
he was never to know, contracted typhoid fever and died
at the age of ten. Their mother, Sarah, also went down
with the disease and for weeks battled with death. She
won, but the grim struggle had taken its toll: she was
partially deaf, and as she grew older her handicaps
became progressively worse. But she was not, as those
who crossed her (including her as yet unborn son) discovered, mentally impaired.

Because of their daughter's death both George Franklin
Getty, an insurance lawyer, and Sarah Catherine McPherson Risher Getty, a former schoolteacher, were perhaps
emotionally ill-equipped to become parents again. Nevertheless, Sarah soon became pregnant and, at the age of
forty, gave birth to Jean Paul on 15 December 1892. Her
husband George, three years younger than her, at the
time of his son's birth was aged thirty-seven.

Sarah was determined that this child would not be
allowed to die. He would be guarded from the possibility
of infection by a possessive, overprotective mother, yet
Jean Paul was never given any physical affection. 'He
was not a child that was cuddled . . . His mother was
pleased with her accomplishment at delivering this boy,'

says Getty's fourth wife, Ann, 'but maybe losing his sister made her afraid to love him. Paul never had a birthday party. He never had a Christmas tree, so he told me. He wasn't allowed to play games with most of the boys because they [his parents] were afraid he would get hurt. Still, he did stand in great awe of them, and I think he loved them.' His parents were strict, puritanical Methodists, rejecting alcohol, tobacco, dancing and the theatre – things that would be dear to their son later in his life.

Some mystery surrounds Getty's birth. On the official registration card only the name of his father was entered; the spaces for the names of both mother and child were left blank. The initial record of the birth was filed in the normal way by the midwife – in this case Dr Mary Whetstone, one of the first lady doctors in Minneapolis. That the parents had not decided on a name for their son is possible but unlikely. But what conceivable reason was there for omitting Sarah's name? Neither parent ever officially registered the birth. It was not until fifty years later, in 1942 and after Sarah had died, that her favourite cousin entered J. Paul Getty's name on the original birth certificate.

His first name was both unusual and a puzzling choice for the son of parents of Scottish-Irish origin. As Jean in English is a girl's name, presumably his parents meant it as the French for John. This happened to be the name of George's strikingly handsome cousin and law partner until the association broke up in 1894, two years after Getty's birth. That there was some strong emotional bond between Sarah and John there can be no doubt. Twenty years later, when John committed suicide, Sarah became an emotional wreck.

After the birth of their son, George and Sarah apparently grew increasingly apart both physically and emotionally. The Minneapolis city directory for 1892–3 gives

George F. Getty's address not as the matrimonial home, but as the West Hotel, a sombre six-storey gothic building. In 1896, when Getty was only four years old, his father almost died from typhoid. He pulled through, but turned his back on his earlier Methodism and became a Christian Scientist, thereby causing a rift with his wife. Young Getty must have realized at some point in his life that there was trouble between his parents because he later wrote to his father, while admonishing him not to interfere in his own marital problems, that 'You know, you yourself, at one time, were going to get a divorce from mother.'

The uneasy relationship between George and Sarah had not always been so. In courtship and early marriage they had been devoted to each other and their son tried to perpetuate the myth that they remained so. Of them he wrote in his autobiography: 'My parents were an ideally balanced couple and devoted to each other. Theirs had been a partnership in every equal-sharing, two-way sense of the word ever since they met.' Yet in 1908 a picture was taken which is perhaps symbolically revealing. George Getty is contentedly draping his left arm over the shoulder of his attractive, lively niece, June Hamilton. Sarah stands alone on the far side, a dour, overweight figure, looking sternly and disapprovingly at her husband and his niece. The sixteen-year-old Jean Paul stands by himself in the centre of the picture, separated from both parents, already a handsome young man, but with a pouting expression.

His parents had met at a small university which had been founded by the United Methodist Church for poor students in Ada, a little farming community in Ohio. It was here that George Getty learned 'not so much algebra, Latin, mental philosophy or any other particular study which was on the curriculum: but the deepest lessons

which I learned were work, sympathy, human kindness.'
They were married on 20 October 1879. Sarah was
twenty-seven, George twenty-four. They were to follow
the great American dream westwards, first to Michigan
for law school and some judicial clerking, then across the
Great Lakes to Minneapolis, and finally via Oklahoma to
California.

Both came from early American immigrant stock which
was fiercely independent, nonconformist, adventurous.
Sarah's Scottish ancestors belonged to the McPherson
clan, George was Scottish-Irish, the direct descendant of a
Presbyterian family which emigrated in the late eighteenth
century from the village of Cullavmor, in what is now
Londonderry. His grandfather, John Getty, landed in
America around 1790, settling in Maryland. In an attempt
to glorify his family's heritage, Jean Paul Getty, in later
years, would claim that his forebears had founded Gettysburg, Pennsylvania, scene of one of the decisive battles
of the American Civil War where Lincoln delivered the
famous address which freed the slaves. Nothing could be
further from the truth. Gettysburg was founded by the
Gettys family, not Getty.

Perhaps the eccentric strain in the Getty family surfaced
early on because John Getty, Getty's grandfather, was
considered a 'lonely, somewhat embittered Irish gentleman, little understood by the frontier folk around him,'
according to one of the family members. He became a
tavern owner, fell off his horse when drunk and froze to
death. His son died of diphtheria when George Franklin,
J. Paul's father, was six years old. He was forced to
become an errand boy, then a farmhand to supplement
the family's meagre income. As he grew older, George
Getty's consuming fear was the poverty he had known in
boyhood. He worked hard and rose to become the chief
lawyer and director of the Northwestern National Life

Insurance Company. The family frequently changed houses in Minneapolis as their situation improved. Jean Paul experienced six of these moves during the first thirteen years of his life.

Minneapolis is almost exactly in the middle of the North American continent in the 'North Star' state of Minnesota and lies mostly on the west bank of the Mississippi River. In 1883 it already had a teeming population of 150,000 and was growing fast. Many emigrants from Scandinavia had been drawn there because of the climate and the landscape – lakes and pine forests – which resembled their own home. But Minneapolis developed rapidly as a lumber, agricultural and milling centre.

From his window in the Northwestern Life Insurance building George Getty could look out on the prosperity of the town, dominated by the four-storey offices of the Western Union Telegraph Company. The avenues, with their stone pavements and mud streets, were crammed with horse-drawn buggies and lined with fourteen-runged telegraph poles, their wires all leading to Western Union.

At the turn of the century, at the age of forty-five, George Getty had an eight-year-old son, a wife and, thrift being his motto, a substantial bank account. He was a successful businessman, not known to speculate, hard-working, a pillar of the community. Had it not been for a bad loan of $2500 that was owed to his insurance company, the story of the Gettys might have ended there. Instead, in 1903, George Getty travelled night and day by train across the farming state of Iowa, through Kansas cattle country to Bartlesville, Oklahoma, to try to collect the debt.

Oklahoma was a territory dominated by Indian tribes. It was still four years away from being admitted to the union, and two years away from the Glenn Pool oil strike of 1905, which would open up the huge midcontinental oilfield. When George Getty arrived the wildcatters were already digging in northern Oklahoma. Bartlesville, close to the Kansas state line, was a small frontier boom town, its mainly Indian population swelled by grubby outsiders in search of the big strike.

The Indians had known of the black stuff for decades. It seeped through rock fissures, often in pools and streams. They believed it to be a magical cure for all sorts of ailments, including rheumatism. The white man first drilled for oil around Bartlesville in 1870, but it was not until a gusher blew in April 1897 (the Nellie Johnstone No. 1 well on the Caney River) that the oil-producing era of Bartlesville and Oklahoma began.

News of the find travelled like wildfire and within five years the town was swamped with oilmen and derricks. Each day the railroad brought in hundreds of hopefuls and the town quickly spread from a few established buildings to a sea of ramshackle huts and cabins. The newcomers included the usual human flotsam of hookers, hustlers and hoodlums. Bartlesville was ready for them. It had a history of lawlessness. Although liquor was banned in Indian Territory, Bartlesville consumed $156,000's worth a year and the town was full of young ladies in rooming houses with no visible means of support. In the town's only poolroom, oilmen and bankers rubbed shoulders with outlaws. It was the haunt of Henry Starr, who boasted that he had robbed more banks than anyone else – until he robbed one too many and was shot dead in Harrison, Arkansas, in 1919. It was also home to Emmett Dalton, one of the infamous Dalton gang who crossed the Kansas border in 1892 to rob the bank in Coffeyville. Only Emmett survived; his two brothers and their two companions were slain.

Into this lawless maelstrom in 1903 stepped the devout

Christian Scientist and stiff-collared insurance businessman, George Getty, who had probably never even seen a
gun, let alone carried one. It is not known whether he
collected the debt or even investigated it. All that is
certain is that, on that first visit, the insurance man
became George Getty the oilman.

Oil represented America's future, a print-your-own-
money business for the successful wildcatter. Henceforth,
the economy would be driven by oil, not coal or steam.
All the signs were there. In 1903 came America's first
coast-to-coast crossing by motorcar (even if it did take
sixty-five days); Orville and Wilbur Wright flew the first
powered aeroplane; motor taxis appeared on city streets;
in Detroit an aggressive entrepreneur with just $100,000
founded a small company to build motorcars – the Ford
Motor Company.

George Getty could not help getting a severe case of oil
fever. For $500 he obtained an 1100-acre tract in the Osage
Nation Territory to the west of Bartlesville. After the first
well was brought in in January 1904, George Getty decided
to bring his family from Minneapolis to the oil-boom town.
At the age of eleven, J. Paul Getty began keeping a diary.
He recorded his first fascination with 'the challenge and
adventure inherent in field operations, in the hunt, the
exploration and drilling, for oil.' He also made his very first
investment – 100 shares of the Minnehoma Oil Company,
the name combining that of the Gettys' former home and
the new source of their fortune. In his first autobiography,
My Life and Fortunes, Getty records his father telling him,
'Now you're part owner of the company for which I work.
You're one of my bosses.' The ownership of shares in his
father's company was to become the source of a bitter
dispute between them and a deep motivating force throughout Getty's life.

Within two months six wells had been drilled, each

capable of producing 10,000 barrels a month. George might have had oil fever, but he remained hard-bitten and stingy. One day Frank Finney, who worked for the Indian Territory Company, took him out into Osage County to look over a lease. It was rough country and hard going. The trek to the spot was half a day's ride in a horse-drawn buggy. When they arrived, George marked the location on a dusty map and headed back to Bartlesville. He did not offer his companion a dime for the day's ride. Many years later Frank Finney quipped to his son, 'I decided right there and then, he'd wind up a rich man.'

It may be that his father's drive for success contributed to Getty's feeling throughout his childhood that he was a loner. His only companion, and the only living creature for whom he ever showed genuine affection, was a mongrel called Jip. He was a belligerent schoolboy, often bringing down the wrath of his teachers, bright but not particularly assiduous in the classroom. In Minneapolis, shortly before moving to Bartlesville at the age of eleven, he almost failed his exams. More effort was put into his leisure time. When he was not taking evening strolls with Jip alongside the dusty railroad tracks around Bartlesville, he would be in his room reading avidly or writing the diary which he began soon after moving to Oklahoma.

His choice of literature is revealing, his diaries even more so. His nightly scrawlings were full of 'Papa goes away, Papa comes back'. He records that his father listened to his piano playing and 'won't let me have any book this week, as he says I read too much. In the evening I read.' The diaries show, even at that tender age, great interest and curiosity in his father's business, for he often records visits to the oilfields. What they do not show is any emotional tie with his mother. She is seldom mentioned and when she is it is only in passing: 'Jip is going with us and also Mama.' He was often left

alone with 'Mama' in the evenings, which he clearly did
not relish. Apart from her overpowering possessiveness,
he probably found it a burden to communicate with the
deaf woman.

An escape from his dull and cheerless childhood was
provided by the novels of G. A. Henty, the English
Victorian author of some eighty adventure yarns for
schoolboys. Getty, like many adolescents of the time,
found Henty's motto for success a simple model of
behaviour to emulate. 'If a lad from the first makes up
his mind to do three things – to work, to save, and to
learn – he can rise in the world,' Henty said. Henty's
novels bulged with the kind of scrupulous attention to
detail that obsessed Getty all his life. They were no strain
to read either, and full of heroic adventures in which
often a ruse or a disguise won the day. Rereading Henty
was a pastime which stayed with Getty until the day he
died. He saw in Henty's heroes qualities that he admired
– virtue, honesty, chastity, heroism. No Henty hero was
ever capable of doing anything small-minded, mean or
despicable. Unfortunately, Getty himself was never able
to live up to these ideals.

Unlike the families of other midcontinent oil barons,
the Gettys did not stay in Oklahoma very long. Sarah
had relatives in sunny California, where the climate would
be far more beneficial for her health. When Paul was
fourteen years old they were on the move again, first to
San Diego, California, and then to the corner of South
Kingsley and Wilshire Boulevard, which was later to
become a high-class address in Los Angeles but in 1906
was little more than a dirt track outside the city boundary.
The Gettys' move to California was made at the time of
the San Francisco earthquake, a natural phenomenon
that was to fascinate J. Paul Getty all his life. George

Getty, one of the largest independent oilmen in Oklahoma, built a Tudor mansion with velvet-lined rooms and tapestries. He had become in three short years a rich man who kept a Japanese cook, a chauffeur, a gardener and had several impressive limousines behind the mansion. The move to the West Coast in effect meant that George Getty would be away from home for considerable periods of time.

Southern California, with its tropical climate, plentiful sunshine and low humidity attracted people either for the scenery or for their health. In 1906 the city of Los Angeles, with a population of almost 200,000, was expanding, thanks to the developing oil industry. Yet within the city limits there were still acres of orange, lemon, walnut and almond groves and vineyards.

The California climate might have been better for Sarah Getty's health, but it did not improve her son's academic achievements. He always claimed his ambition was to become a writer or a diplomat, but despite his autobiographical claim that 'I applied myself to the task and obtained good marks', his school records tell a different story. His father took him out of the John E. Francis Polytech High School because his marks were not up to scratch. He was sent to Harvard Military Academy to be disciplined, but the academy was no more successful than his previous schools, and soon he was to start on his round of universities. He scorned the 'playground' atmosphere at his first, the University of Southern California, and soon switched to the University of California at Berkeley because, he said, it offered courses in both political science and economics. Whatever his ostensible reason for moving north to Berkeley, several hundred miles away from his family, it was not to study political science or economics. During the spring term of 1912 he took seven academic courses: military science, Latin, philosophy, physiology, English, Greek and French. There was no trace of economics or political science. In fact, he failed to finish a single course during that spring term. In a letter dated 2 January 1912 to his parents, he claimed to be studying mathematics, which was simply not true. They might have seen their son was pulling the wool over their eyes, for at one point he said he had been practising the piano, and then, apparently forgetting this fib, he asked his father at the end of the letter to send $42.50 because 'my music and piano are due today'.

In April Getty temporarily took leave of absence to travel for two months in the Far East. The trip to Japan and China gave him 'the first faint stirring of a desire to own works of art', and he was impressed by the polite manners and industriousness of the Japanese. He returned for the fall semester and again avoided all economic and political science courses. His stay was short, and he was 'honorably dismissed', which means he left voluntarily. To save face with his parents, he now arranged suddenly to leave America for Oxford University, which, he claimed, offered wider, less parochial courses in economics and political science – the only two subjects he always said he wanted to study but never bothered to master. He assured his father that, after Oxford, he would study law.

He entered Oxford as a non-collegiate student, a member of St Catherine's Society. Although the Society was not yet a fully accredited college, his tutor was from Magdalen, one of Oxford's most prestigious and socially acceptable colleges. The first term was nearly over when Getty arrived.

As the geographical distance between Getty and his parents widened, so did the emotional rift. George Getty had doubts that his son seriously intended to buckle down to study at Oxford. As it turned out, his fears were well founded. By the winter of 1913 Getty had caught European wanderlust. He roamed the Continent in a second-hand Mercedes-Benz. His parents complained that they often had no address for him other than care of American Express or Thomas Cook.

Letters between father and son grew increasingly acrimonious. The bitterness was focused in the younger man's high, wild life and the cost of supporting it. Each month George Getty sent his son $175, not an inconsiderable sum in those days. But, for the young Getty, it was insufficient. He constantly wired his father with requests for more cash. His parents complained of his life of 'gaiety' on the Continent. 'Neither myself nor your mother approve of the style of living you seem to have adopted,' wrote George Getty.

While he was gallivanting around Europe, a drama was unfolding back home which was seriously to affect the health of his mother. George's personable cousin, John Getty, had spent his life getting involved in one get-rich-quick scheme after another. Each one left him poorer, and the final ones, in real estate and lumber, left him bankrupt. In July 1913 he gassed himself in Fresno, California. It was Sarah Getty, rather than John's wife, who took care of all the funeral arrangements, and after the service she sank into a deep depression. To his son, George Getty wrote: 'She lost consciousness and memory of the whole affair . . . She talked incoherently for days about you and me, thus apparently thinking which one or the other of us had done the deed.' She was ill for weeks and, even afterwards, suffered lapses of memory. Her reaction to the death of her husband's cousin was more akin to the grief that a woman might show for the loss of a mate. Mystery surrounds Sarah's collapse at John Getty's death. George Getty gave no hint of the reason for her emotional state in the letter to his son. Nowhere else in the Getty family literature is this episode mentioned, but Sarah Getty remained so ill she was unable to visit Europe that summer.

If Getty was concerned about his mother, he never showed it. There is no trace of any letters home on the subject, but he told his father that he had refused to pay a fee for his Oxford diploma. A few days after Sarah's breakdown he received a piece of sound advice from George: 'I would think you would want to get hold of it [the diploma] – even if you have to pay some fees which you dispute. The authorities know what is customary and they have a big advantage over you – even I suppose of indefinitely or permanently withholding the diploma in case you fail to comply with the requirements. You had better get hold of it, if you have not already done so.'

It is unclear whether Getty ever acquired his Oxford diploma. In light of his earlier unstable academic record, he may never have completed the necessary work and the complaint about paying for the diploma might well have been his way of concealing the fact. However, the college records support his claim of a diploma, although the current master says that 'there may well have been no grades' required for a diploma. Curiously, his autobiography states that he spent two years at Oxford and received the diploma in 1914 rather than June 1913 as recorded by Oxford. Like Jay Gatsby, Fitzgerald's fictional hero, Getty made good use of the Oxford diploma. It gave him a certain cachet. In later years the slender Magdalen connection became part of his masquerade to gain social acceptance in Britain – he claimed to have made friends there with the Prince of Wales.

In fact, Getty was more concerned with the subject which was always to dominate his life – money. George, despairing that his son would ever put his life in order, wrote to tell him that he was seizing the 15,000 shares in his son's name in the family's Minnehoma Oil Company. George had had a sports car built for his son in California, and he wrote: 'Your stock was more than eaten up in the expenses of building that automobile.' Thus, only weeks away from his twenty-first birthday and legal majority, J. Paul Getty discovered that his one and only investment now belonged to his father.

His reaction was expressed in a long and venomous letter from Vienna, which amounted to a declaration of war. It is a revealing document because it laid down clearly the ground rules not simply for Getty's future dealings in business but also in his private life. He argued strongly that the $2250 in Minnehoma dividends had more than paid for the car. He bitterly demanded the Minnehoma shares, his automobiles (there were two in all) and some $450 in cash. He wrote:

Many fathers in your position would have given their only son this as a nest egg . . . When Ned Haupt was 21 his father gave him $125,000 worth of Los Angeles real estate. When W. R. Hearst was 21, his father gave him the *San Francisco Examiner* and the *Examiner* building, valued at $3m. So that it cannot be said that what I will come into is very large. Yet you try your best to make it nothing at all, and if I were feeble minded, I might believe you and so be cheated out of my birthright.

He continued: 'I admit I haven't very much coming to me, but your "absolutely nothing", my dear father, is a little too much', and he characterized his own letter as one written 'by a lawyer to a Jew money-lender, not from an only son to his affectionate and generous father.' The gauntlet was thrown down. 'Your express attitude leaves me no choice. No choice but to deal with the matter as though I were dealing with an opponent.'

The young man then proceeded to attack his father's business acumen, criticizing him for not building storage tanks in Oklahoma, which Paul had earlier advised. 'So, by wantonly disregarding my advice, and flaunting all the rules of business and economics, you have cost your stockholders . . . enough to pay a 5% dividend.' Even so, there followed a few more tips on running the oil business which are inconsistent with the business morals Getty was to follow. He told his father to bribe the Indian Council with $600 for drilling rights on a lease which was in dispute between Standard Oil of New Jersey and George Getty's company. To block Standard Oil, the letter suggested, 'a good plan would be to get an injunction against their drilling till the case is decided and meanwhile drill along their line, tapping their pool.' George Getty, who in his business life at least had a reputation for straight dealing, replied: 'I had rather have the $600 and my honesty and honor.'

In June 1914 George and Sarah Getty took a trip to Europe, a visit which had been earlier postponed because of Sarah's illness and George's business commitments. George wrote to his son from Oklahoma, telling him of their plans and asking him to meet them in Hamburg. Sarah, who had written a long and affectionate letter ('I think of you constantly and sometimes wish I could fly to you'), was longing to see her son, but he failed to turn up. After a separation of a year and a half, George and Sarah Getty were disappointed and angry that their son was not there to meet them.

When they did meet, the dark rumblings of the First World War were already audible. Despite his wide-ranging travels around Europe, however, young Paul Getty had not the slightest suspicion that war was coming to the Continent or bloody revolution to Russia. The family were all together when war broke out. George Getty kept a diary of the trip and, on 30 August 1914, shortly after hostilities began, he wrote: 'Sightseeing has lost its interest in the contemplation of these awful things impending. The important thing is to get away from it all and to forget its horrors. The anger, hate, revenge, malice, mad ambition have been reduced to their native nothingness.' He added, with baseless optimism: 'We shall have peace, lasting peace.'

The Gettys returned to the United States on board the liner *Lusitania*, which less than a year later was torpedoed off the Irish coast with the loss of 1195 lives, bringing the United States into the war.

Getty claimed years later in his autobiography that, upon his return to the States and before he left New York, he had volunteered as a pilot or an artilleryman in the American forces. Whether he volunteered or not, he remained a civilian. Mysteriously, Getty, a fine physical specimen, was never called up, even though many of his peers were ultimately inducted into the services.

On the journey home Getty thought about his future. He insists in his autobiographies that his goal was to become a diplomat and a writer. But his academic career had been a shambles and he showed no interest in following his father by training as a lawyer. More likely, the younger Getty had an inner drive to make money, to retrieve his 15,000 shares of Minnehoma and more. He agreed to try his hand at the oil business. George Getty would advance him $100 a month, about half his living expenses in Europe, and provide funds to buy 'low-cost leases on properties I consider promising'. He would share the profits – 70 per cent to George Getty, 30 per cent to himself.

George was philosophical. 'If the experiment doesn't work out or you are unhappy when the year is over, you can do whatever you wish. I won't say another word.' He accordingly cut his son's allowance; the young man, without another nickel to his name, left for Tulsa in Oklahoma where a billion-dollar poker game was in progress.

3

The First Million Dollars

Oil is like a wild animal. Whoever captures it has it.

J. Paul Getty

Throngs of fortune hunters – hoboes, gamblers, the inevitable prostitutes and hard-bitten wildcatters who had already tried their luck in the fields of Pennsylvania and Illinois – were in Tulsa when Getty arrived there in 1914. Some would make and lose fortunes overnight and never be heard of again. Others would become millionaires. Rough-hewn Bill Skelly, a mule driver from Pennsylvania, and Frank Phillips, a tall, bespectacled Iowan who looked like a preacher, would start and build major oil empires that would be called Skelly Oil and Phillips Petroleum.

Most of these men were self-made, uneducated and unlikely to have spent much time in New York, much less the sophisticated capitals of Europe that were Getty's old haunts. Yet some of them became his mentors, among them P.M. McFarlin, part owner of the McMan Oil Company. Production in Oklahoma had soared from virtually nothing in 1900 to 74 million barrels in 1914. At a price of 78 cents a barrel, this new American industry produced revenues in Oklahoma of $58 million in that year alone.

Into the raw and ruthless world of the Oklahoma oilfields came the self-willed son of George Getty, not quite twenty-two years of age, the only man in Tulsa to wear a wristwatch. Many of the wildcatters carried guns instead. Tulsa, with the railroad running down its main street, was growing fast but, with its mud thoroughfares, was hardly a civilized place to live, and outside the town life was even tougher. Oilmen spent their days ankle deep in the slime churned up by the horse teams and supply wagons and their nights in makeshift, draughty wooden shacks that served as bedroom, kitchen and cupboard. What there was of food tasted like garbage from a hotel dustbin and the stench of unwashed bodies was blotted out only by the oblivion found in alcohol. The occasional visits to nearby towns were only for the purpose of spending the odd hour in a brothel. Getty boarded at the grubby, primitive Cordova Hotel for $6 a week and was forced to take his meals at a greasy shack down the road for another $1 a day.

He had a reservoir of practical knowledge about the oilfields. For several summers he had worked on his father's oil rigs, sharpening and repairing drill bits. This experience had toughened him physically and helped him develop a rapport with oilfield workers and roughneck wildcatters. 'I'm still a wildcatter at heart,' he liked to say. It was an expression of the spirit of the loner, the unorganization man, who never wanted to be ordered about or tied down to a routine. It must have been infuriating for him to be so firmly under his father's control. George Getty laid down the maximum that his son could bid for any single lease, with the result that he was forced to deal only in low-priced oil and gas leases which were avoided by more daring speculators. He wrote many years later in the magazine *Playboy*: 'My first year was anything but profitable. Large oil strikes were being reported regularly and other wildcatters were bringing in gushers and big producers, but fortune seemed to elude me.'

One place to look for oil was close to an area where a strike had already been made, for example, in Stonebluff, a small town in the rolling sandstone hills some 30 miles south of Tulsa. It was here that Getty made his first killing, purchasing in late 1915 a half interest in a lease on the farm of one Nancy Taylor.

Years later Getty used to enjoy telling how he made his first oil strike. He would recount how he set out one morning in his third-hand Model T Ford, a less noble vehicle than the one he had been used to in Europe, and drove south from Tulsa along the dusty country roads. They led through a stark, spartan landscape, straight through Stonebluff, where the gang of Tulsa oil traders had already staked out some valuable holdings. Getty's destination was a little farther south, on the way to Haskell, a couple of miles from a bend in the Arkansas River. He wanted to see Nancy Taylor's farm at first hand. It was situated on the highest ground for miles around, spattered with the occasional regiments of oak, elm and maple amid the coarse grass and sumach that bound together the harsh soil of northeastern Oklahoma.

But it was not only Getty who was interested in the farm. Other wealthier independent oilmen wanted the lease just as badly and might bid up to $15,000, way beyond the limit which his father had laid down. He dared not force a confrontation with George but he had no cash of his own with which to bid, so he set about putting the other contenders out of the running by employing a crafty ruse. He drove to the small county town of Muskogee, where he persuaded a local bank official to bid for the lease on his behalf. His reasoning was clear. Some independent oilmen who owed the bank money would be reluctant to bid against it. Others would assume that the bank was acting for a major oil company with more financial muscle than all the independents put together. The ruse worked. Getty secured his interest in the lease for a mere $500, the same price his father had paid in 1903 for the 1100-acre lease near Bartlesville that was to become the foundation stone of the Getty family fortune.

That Getty struck oil in 1916 on the Nancy Taylor lease there can be no doubt. But can his often told story of how he put one over on his rivals he relied on? On 9 February, the day after the well 'came in' (oil language for 'began producing'), the *Tulsa World* reported that 'with all the excitement at Stonebluff, this neck of the woods has been overlooked', adding that it 'was certainly a surprise to the [oil] fraternity'. If, as Getty said, other independents had been interested in the Nancy Taylor farm, it could hardly have been such a surprise when he struck oil. It made an amusing and interesting after-dinner story but it may be no more than that. Indeed, he himself admitted later that 'a lot of people thought there was no oil there; and even more people tried to persuade me that I was in the wrong place to drill. But I was stubborn and not about to be moved. It paid off.'

Getty sold the Nancy Taylor lease to Cosden Oil and Gas, one of the first refiners in the Cushing field. Josh Cosden was one of the new super-millionaires in Oklahoma and entertained the Prince of Wales in his Tulsa home.

J. Paul's 30 per cent share of the profit from the sale on 12 February, three days after the strike, was $11,850. It was the young man's first profitable oil deal, but it was a landmark for another reason. Because of it, George Getty formally accepted his son into the family business. Getty was made a director of his father's master company, Minnehoma, which owned leases in the Cushing-Drumright field some 9 miles west of Bartlesville where, during 1916, some 17 per cent of all the oil in America was produced. True, Minnehoma was a minor member of the Oklahoma oil industry. By 1916 it was producing only a few hundred barrels of oil a day and was worth somewhere between $1 million and $2 million.

In those early years the company's fortunes were subject to the ebb and flow of crude-oil prices, which between 1903 and 1916 swung wildly between 30 cents and $1.20 a barrel. In times of an oil glut, small independents like Minnehoma were at the mercy of prices established by the owner of the only pipeline from the midcontinent to markets on the eastern seaboard – Prairie Oil and Gas, which had once belonged to Standard Oil and still retained close links with it after the monopoly had been broken in 1911.

However, the posted price of oil, which fell from $1.40 a barrel in April 1914 to 40 cents a barrel in February 1915, soon rose to $1.20 a barrel in 1916 and to $1.73 in 1917, because of the First World War, which was fought by mechanized transport both on the ground and in the air. In 1918 it reached a high of $2.23. When Lord Curzon said, 'The Allies floated to victory on a sea of oil,' few people realized that one of the new fortunes being made in the fields of Oklahoma belonged to George and J. Paul Getty. In a way, young Getty, whose application to join the armed services had been rejected, could well imagine he fulfilled his patriotic duty in helping to produce the crude oil that fuelled the tanks and trucks at the front in Europe.

Getty's success at Stonebluff was the beginning of a continuing business association with his father. Several new Getty oil enterprises were eventually created, including, in May 1916, the first Getty Oil Company – then a much smaller, less important entity than Minnehoma or the California-based George F. Getty Inc., which was founded in the early twenties. Not until 1956, four decades later, would the parent company of the oil empire be called Getty Oil.

During this period Getty continued to live at the Cordova Hotel with a group of other wildcatters, but he spent more time in the lobby of the Hotel Tulsa. When it was empty, its wide, spacious lounge looked like the reading room of a London club, with shiny, old brown leather armchairs and couches and a fan-shaped marble staircase. But, during the day, it was more like the New York Stock Exchange, teeming with oilmen picking up information on oil prices, recent strikes, likely prospects, who was drilling where, and on the buying and selling of leases. Between 1912 and 1927 $1000 million's worth of business was done in the hotel.

It was there that Getty learned very quickly, as he revealed in an interview, 'that the one who had the most information was the one likely to get the most information in return . . . After my experiences in Tulsa I knew all the nuts and bolts of the oil industry.' He later wrote in his autobiography, *My Life and Fortunes*: 'My leasebrokering business boomed, and I made sizable profits on a large number of transactions. I also acquired several leases on properties I intended to develop.' He steered clear of the established oil-producing areas of Oklahoma, for he had neither the patience nor the cash to compete with the older, giant independents. So he concentrated on those areas shunned by the heavyweights, such as Noble and Garfield counties to the north and west of the pulsating Cushing-Drumright field.

Most oilmen scorned the idea that beneath the red beds of these counties lay something besides red-brick shales and clays. Getty felt that, 'as with all things in nature, there must be some logical order to the manner in which petroleum was distributed beneath the earth's surface.' And so he became one of the few independent oilmen to use the relatively young science of geology. Many of the old-timers did not trust the new science.

After all, most of Oklahoma's oil had been discovered from the telltale signs on the surface – crude oil seeping into streams or through fissures in outcrops of rock. Yet, in 1913, three years before the Nancy Taylor strike, Charles N. Gould, the founder of the Oklahoma Geological Survey, had presented a paper at the International Geological Congress in Toronto, in which he cautiously concluded: 'Careful studies of geological conditions have demonstrated that there is a rather definite relation between structures of rocks and occurrences of oil and gas.'

The first geological department established by an oil company to work in the midcontinental area was Union des Petroles, set up in Holland with French backing. It sent several Swiss geologists to the USA including Emil Kluth, who became the chief geologist for the Getty family. Some of Kluth's earliest surveys were in the area where the Nancy Taylor plot was located. Kluth's boss, Edward Bloesch, found that 'the oilmen were ignorant about the value of geology, considered it a fad and made jokes about it.' A rare exception was the former Oxford student, J. Paul Getty, who used Bloesch's studies to discover the Garber and Billings fields which were located a few miles west of accepted oil territory in northern Oklahoma.

Geology apart, there were good historical reasons for believing that oil lay under the red earth of Noble and Garfield. In 1911, one of the industry's most famous wildcatters, Ernest Whitworth Marland, who eighteen years later would take over the Continental Oil Company, had discovered and opened up the giant Ponca City field just north of Noble County.

Getty's geologist was right. Between February and June 1916 Getty indulged in a frenzy of lease buying and selling, of drilling and producing, of wheeling and dealing.

After a mere five months he 'suddenly discovered' that he was worth $1 million. Or so he claimed. By comparison, it had taken George Getty thirteen years, from 1903 to 1915, to make $426,000 out of the production from Lot 50, his first oil strike in Oklahoma. When production declined to seventy-eight barrels a day, he sold the property for $120,000 in cash – not a bad transaction, although in later years his son would criticize the sale on the grounds that it took place just before the golden age of Oklahoma oil, when prices rose from $1.20 to $3.25 a barrel. In 1916, George Getty, who was president, treasurer and general manager of Minnehoma Oil Company, paid himself a salary of $5000 and collected dividends of $18,543. It would be interesting to know how his son, in the course of a few months, contrived to convert his profit of $11,850 on the Nancy Taylor lease into a $1 million nest egg. Unfortunately, he was remarkably vague on that point. Those months, he said, 'were extremely fortunate ones for me. In most instances, the leases I bought were sold at a profit and when I drilled on property, I struck oil more often than not.' It is true that, by now, he owned 30 per cent of Minnehoma's undeveloped oil leases, which had risen in value four times from their cost of $25,000, but that only amounted to $100,000.

Still, despite the use of geology, wildcatter Getty believed that the most important factor in bringing in a producing well 'was often just plain luck'. He also believed in the law of capture, the philosophy that to the aggressive hunter of oil belong the spoils. As an independent, he familiarized himself with 'all aspects of our business and kept our costs down by exercising unceasing and vigilant supervision over every phase of our operations.' Here was the source of Getty the wildcatter, the renegade who refused to be an organization man. He found his true métier in being a loner.

Whether Getty had made $1 million during the year or not, one would have expected him to have poured the profits he undoubtedly made back into the business in the hope of turning them into an even greater fortune. Instead, he deserted the Oklahoma oilfields and returned to California. At the ripe age of twenty-four he decided to retire and have fun. 'I had never been interested in having a lot of money,' he was to write somewhat disingenuously later, 'and never wanted to be leader of the pack. I always preferred to be a member of it.' All through his life he contradicted himself on the subject of money. During those early years in Oklahoma and ever afterwards, as he told one interviewer, 'I have always been afraid of failure. Always afraid of poverty. I inherited it.'

His decision to 'retire', however, is fairly typical of the lifelong seesaw between a puritanical work ethic and an unadulterated hedonism which was to be the constant rhythm of his life. 'Now I found I'd made enough money to meet any present requirements I might conceivably have in the foreseeable future. I made a headstrong snap decision to forget all about work thereafter and to concentrate on playing, on enjoying myself,' he wrote later in *Playboy*.

What was rather surprising, however, was that when he returned to California he moved back into the family home at 647 South Kingsley Drive in Los Angeles. It was there that a certain Miss Elsie M. Eckstrom asserted that Getty had deprived her of her virginity when she was drunk. She made this allegation in a lawsuit she brought against him in which she claimed that he was the father of her daughter.

This episode with Elsie Eckstrom was the only alleged instance of Getty bringing a woman back to his family's home for sexual purposes. However, Getty had his own entrance and separate apartment at home, enabling him to enter and leave without his mother or father meeting his guests. Miss Eckstrom's story, as filed with the Los Angeles Superior Court, was that Getty plied her with liquor on 3 February 1917 in the downstairs living room until she finally passed out. When she came to she was in bed with him.

Getty, aged twenty-four at the time (Miss Eckstrom was twenty-one), denied the charge. He refused to admit that he had even offered her a drink, far less that she was 'lying alongside of and in the bed of the defendant.' He further claimed that Elsie was 'not a woman of chastity and virtue', but a woman of 'loose character and a constant frequenter of road houses, remaining in said road houses until three or four o'clock in the morning . . . habitually drank liquor to excess, and . . . was frequently intoxicated.'

Elsie Eckstrom's baby girl was not born until 6 December 1917, ten months and three days after the alleged seduction. This did not stop the *Los Angeles Times* from using the headline, 'Baby Born to Getty's Accuser'. The story said that an entry in the city's birth records made the simple announcement: 'Born to Mr and Mrs Paul Getty, a daughter, at the California hospital,' and added that the child's name was Paula Getty. However, the California birth certificate does not give a name for the child, although it lists Paul Getty as the father. His age was incorrectly given as twenty-seven and his birthplace as Oklahoma. By the time the *Los Angeles Times* interviewed him he had taken refuge at the Los Angeles Athletic Club, and there he told the reporter: 'If the child was mine I certainly would have the manhood to arrange for its care and future.'

Clearly Gesner Williams, Eckstrom's lawyer, meant to gouge the new oil millionaire. He told the *Los Angeles*

Times: 'Miss Eckstrom, according to her present intentions, will ask Getty for an amount of money sufficient to care for and educate the child. Unless the settlement is arranged, Miss Eckstrom says she will proceed against Getty under the State laws and hale him before the court for a compulsory settlement. The alternative, if her charges are proved, is jail.'

Two months later Getty and Miss Eckstrom settled what the newspaper called 'their somewhat sensational difference' out of court. It reported that Miss Eckstrom had dropped her $100,000 suit for damages against the 'young man about town' but did not disclose whether there was a cash settlement.

This was the first event in Getty's life to be sensationalized in the press and it must have been mortifying for his parents. For his part, the 'young man about town' decided to come out of retirement and return to the oilfields alongside his father. Perhaps this was in order to win his way back into the good graces of his parents, or perhaps in order to replenish his coffers after the legal expenses of the paternity suit. 'A total lack of work,' he declared, 'brought me nothing but boredom, restlessness, and a sense of futility.'

4

The First Three Wives

What gives a man his drive is sex . . .

J. Paul Getty

When America entered the 1920s, oil was already becoming the source of southern California's wealth. Early Spanish settlers in California had used the sticky black substance found on the beaches to waterproof their canoes and their baskets, and to grease the axles of the carts carrying fresh fruit and vegetables to market. They called it *la brea*, and by 1920 the huge increase in the number of automobile registrations and the dawn of air transport was turning *la brea* into pure gold. One out of every four barrels of oil in the USA was already being produced within a few hours' drive of the Gettys' home. Gushers were everywhere; the most spectacular series of them ran for 43 miles along the beautiful sandy beaches from Los Angeles south towards San Diego.

Getty did not himself discover any of the great fields south of Los Angeles. It was Union Oil that found Santa Fe Springs in 1919 and Standard Oil of California that brought in Huntington Beach in 1920. However, production from Santa Fe Springs and Long Beach was a major boost in 'increasing the Getty fortune', according to Emil Kluth. He later claimed that he pointed out 'the outstanding topographic feature in a line from Huntington Beach to Seal Beach, Long Beach, Dominguez, Athens, Baldwin Hills to Brentwood Country Club,' the location of Twentieth Century Fox's film studios during the 1950s.

Unlike the Oklahoma oilfields, production in California

was dominated by several giant concerns such as the Southern Pacific Company, the railroad that serviced the area, Standard Oil of California and Union Oil, a local company with a major stake. Even so, adventurous wild-catters could make a killing around Los Angeles. All they had to do was buy a lease in one of the coastal towns, where farmers' plots were being turned into derrick cities, and drill close to someone else's well. 'There were no dry wells or holes. It was like shooting fish in a rain barrel,' said Garth Young, a field engineer for Signal Oil and Gas, one of the new independent companies.

In late 1921 the Gettys bought a lease on Telegraph Hill in Santa Fe Springs for a mere $693. They were early enough and shrewd enough to secure a plot of land close to the crest of the hill, the location of the oil pool's dome. The timing was decisive. By 1923 the wooden derricks that covered Telegraph Hill made it look like a pincushion. Production soared from 218,000 barrels in 1921 to over 70 million barrels in 1923. George F. Getty was the president of a new company, George F. Getty Inc., founded in 1924 to develop the California interests, in addition to the modest Minnehoma which was based in Oklahoma. Over the next fifteen years this one lease would pour $6.4 million into the new company. In later years J. Paul Getty was to take the credit for his father's estimated fortune as well as his own. He claimed that George F. Getty Inc. would have been unprofitable during the 1920s had it not been for the property which he personally assigned to his father's company.

The sixty-eight-year-old George Getty and his thirty-year-old son could not have been more different in their style of operation, personality or business values. George's honesty in the oilfields of southern California was legendary. Garth Young remembers that George Getty told him that a signed contract for the delivery of crude oil and gas was unnecessary. 'If I give my word, no one can break it, but if I sign this contract, my lawyers can break it.' He remembers Getty junior as a 'tall, gaunt, gimlet-eyed wild sort of young man', who 'was only interested in getting money in order to get more money'. Indeed, his 1920 passport photo shows Getty with a determined, steely look, the demeanour of a cardsharp on a Mississippi steamboat rather than a polished Oxford graduate.

Young believed that George Getty never trusted his son. Sometimes, when the young man arrived unannounced at the oil rigs at Santa Fe Springs and tried to commandeer equipment for his own drilling, George Getty would remind his men, 'I told you to keep that son-of-a-bitch out of here. He's not to have anything.' At other times he would soften and say, 'Well, he's my only son. Let him have it.' Young did not like the way Getty 'always came to the fields later than his father, rode on his father's name, and rounded up a lot of wealthy people who invested in wells.'

In March 1923 the older man was brought down by a serious stroke and it probably did not surprise him that his son at once tried to assume control of George F. Getty Inc. Despite a partial paralysis that affected his speech and one side of his body, the indignant George Getty ordered his son out of the driver's seat and resumed control from his sick bed. Although he must have been angered by the rebuff, J. Paul Getty later asserted that it was with 'overwhelming joy that I watched him improve slowly as the pressure of the work he took upon himself mounted.'

George Getty certainly had aides who kept watch on his son's business activities, but he may not have been aware of his adventures around Los Angeles with lively teenage girls for whom his admiration amounted almost to an aberration, as if he were afraid of mature, more experienced women.

In 1922 a dancing man could find partners at any hour of the day. There was dancing at the plush Rendezvous Restaurant in Los Angeles where high-spirited fourteen-year-old Ann Rork preferred to spend her lunch hour, and when she returned home from school one day in 1922 there was a message from a 'Mr Paul' asking for a date. This was the pseudonym Getty used in his love affairs. Ann's mother allowed her to go on dates with 'Mr Paul' and twice he took her to the Coconut Grove, then the 'in' night club in Los Angeles. However, Ann's father, Sam Rork, a well-known Hollywood producer and the agent of the silent film star Clara Bow, knew all about 'Mr Paul' and forbade Ann to see him. He knew of Getty's reputation as a member of the fast set in Hollywood where, already in the early 1920s, it was important to be seen with the right sex symbol in the gin joints. Rork must have wanted his daughter to be far away from the man who had made the headlines with Elsie Eckstrom, and he sent her to Knox School, a private boarding school in Cooperstown, New York.

Getty, at the time, was a well-known figure at Jack Doyle's boxing arena in Vernon, outside Los Angeles. He often appeared with fast and beautiful women, according to Teddy Hayes, a Los Angeles neighbour who became Jack Dempsey's trainer. In fact, Hayes claims that Getty sparred three rounds with Dempsey in 1923, and 'he did very well. He was a very good fighter.' Photographs of Getty in a bathing suit show a wiry agile frame, kept in shape by punching a big sandbag in his private basement gymnasium. Dempsey once described Getty as 'well built, pugnacious by nature, and quick. I've never met anybody with such intense concentration and willpower – perhaps more than is good for him. That's the secret.'

While he sparred with boxing champions, his taste in women still ran to appealing high-school girls. Slightly raising his age target, Getty began to court eighteen-year-old Jeanette Demont, whose huge dark eyes reminded him of his mother in her youth, or so he said. He did not introduce Jeanette to her, however, or to his father. Instead, he eloped with her to Ventura and married her there on 1 October 1923. A photograph of the newly-weds shows Jeanette elegantly dressed, her right hand jauntily on her hip, while Getty, scowling into the sun, looks stiff and formal in a double-breasted suit.

When Sarah and George finally met Jeanette they found they had an intelligent, articulate daughter-in-law. They must have been under the illusion that the wild playboy had finally settled down and were even more gratified when, on 9 July 1924, George Franklin Getty 11 was born. They probably were unaware that less than two months after his marriage Getty had begun taking out other women to the Cinderella Roof and the nearby Ambassador Hotel, two of Los Angeles's hotspots. It was to emerge in court that, early in 1924, Getty shouted at his wife in front of friends, 'I'm sick and tired of you . . . sick and tired of being married.' Within less than two years of their marriage, Jeanette filed for divorce. Then, as now, aggrieved parties in a divorce action are apt to exaggerate the most minor marital troubles. Even so, Jeanette's petition, first filed on 20 August 1925, is a catalogue of misery. It tells of physical intimidation, indifference and cruelty. In mid-February 1924, when she was four months pregnant, Jeanette claimed that her husband threatened to kill her. Shortly after the birth he locked her in a closet and walked out of the room. A few months later, when he was fined $100 for being in possession of illicit alcohol at a party at a hideaway in the Hollywood Hills which served as a private bordello, he turned on his wife in his rage and 'beat and bruised' her.

Whatever the truth of Jeanette's courtroom allegations, there is no doubt that by the middle of 1925 she was an emotional wreck. Her doctor had her admitted to the Alvarado Convalescent Home, where she remained throughout May and June. When Getty visited her there he was so abusive that Jeanette suffered severe nervous shock and went into hysterics.

She claimed that her husband had threatened to liquidate his entire oil interests to avoid paying her any alimony, and asked the court to grant her alimony of $1500 a month. She also wanted the court receiver to take possession of all her husband's assets. These pleadings were made under California's community property statute which, in certain circumstances, gives a divorced wife a slice of her husband's property. It was Getty's first brush with this particular law and it was a lesson he took to heart, even though Jeanette's attempts to freeze his assets failed. The court awarded her alimony of $500 a month until December 1936, when the amount would be reduced to $375 a month for six months, and to $250 a month thereafter – even if she remarried.

Despite the bitterness of the divorce, some five years later, on the anniversary of their elopement, Jeanette sent a simple one-word telegram to Getty which showed that she still had tender feelings for him. 'Memories', it said.

Getty left Los Angeles in disgrace in the summer of 1926, planning to enter the oil business in Mexico and Venezuela. Clearly worried about his parents' opinion of him, he wrote to his father from Mexico City on 8 July claiming that Jeanette had lied in her divorce papers. He meant to impress his parents with a determination to

make a success in the Mexican oilfields where already
1.25 billion barrels of crude oil had been produced from
shallow wells. Getty complained throughout the letter of
being separated from his father and wished that his
mother would write to him. While hinting that he would
like to come back to work for his father, Getty says, 'I
seem to be continually urged on by a great ambition to
accomplish things, especially when I am away from Los
Angeles.'

Getty apparently remained in Mexico for six months.
There his impulsiveness and his taste for young girls were
about to betray him once again. He met the teenage
daughters of a Texan rancher, seventeen-year-old Allene
Ashby and her sister Belene. Allene and Getty eloped
to Cuernavaca, a small village nestling under the twin
volcanoes of Popocatepetl and Ixtaccihuatl and there they
were married on 27 October 1926.

In fact, Getty was still legally married to Jeanette.
Despite his later assertion that he had been divorced
from her on 15 February 1925, according to the *Los
Angeles Times* their decree did not become final until 22
September 1927, so his marriage to Allene was bigamous.
At the same time he was having an affair with Belene.
His subsequent wives report that in later years Belene
was often taken along as a third person on a date and
Getty left her money in his will. A third sister, Lois
Jensen, worked for a time for the Getty Oil interests.
Small wonder, then, that within weeks of their marriage
he and Allene decided that they had made a mistake and
separated. Later Getty was to claim that he was divorced
from Allene in 1928, but the exact resolution of this short
marriage remains cloaked in mystery. In Cuernavaca in
the 1920s it was possible to arrange practically anything
with a good local lawyer and some ready cash.

Getty's sexual licence obviously widened the rift with

his parents. Miss Mary Kennedy, a friend of Sarah Getty, says that Sarah felt her son 'had been taken over by the devil'. And even if George Getty, the Christian Scientist, could do nothing to bring about the spiritual salvation of his son, he evidently decided that a little temporal intervention was appropriate. In December 1926 he removed him from the board of Minnehoma Oil and Gas. And, unknown to his son, he drew up a new will.

In June 1927 the thirty-four-year-old adolescent took a trip to Europe with his parents in an attempt to patch things up. The attempt was evidently a failure. Two months later Sarah and George returned to the United States, but Getty remained in Paris where he rented a small bachelor flat at 12 Rue St Didier in the shadow of the Eiffel Tower. For the next twelve years he spent at least five months each year in Europe, despite his business involvements.

In the summer of 1928 he was in Vienna. There, in the Grand Hotel, he met a tall, vivacious and totally innocent teenager, Adolphine Helmle, who had just finished a convent education and was travelling with her parents. He followed them to Venice and asked her to marry him, and then returned to Paris for a few days. But just when she thought he had set sail for America, he turned up in her home town of Karlsruhe. Otto Helmle, her father, a doctor of engineering and head of the Badenwerk electricity company in Karlsruhe, was opposed to his daughter marrying a divorced American more than twice her age, but Adolphine, nicknamed Fini, finally won her parents over. Getty claimed that he too had sought the consent of *his* parents. After two failed marriages he did not want to 'cause them any more emotional upset or pain'.

In December 1928 Getty married Fini in Havana and spent his honeymoon in Florida. This meant that he had been married three times in the space of five years.

Clearly his marital entanglements were more impulsive
than any investment he made in oil wells. Also in
December he was busy trying to renegotiate a prenuptial
agreement to pay Allene $10,000 down to $5000. He
offered to buy her $5000's worth of stock in several listed
companies, but it was to be registered in his name until
she reached the age of twenty-one. To make up for the
loss of $5000, he also offered to pay her $150 a month if
she would let him keep the cash dividends for the stock.
It was a financial strategy that Getty was to employ years
later with the shareholders of companies he wanted to
control. There is no evidence that Allene accepted his
offer.

Never one to be modest about his achievements in the
oilfields, the younger Getty took credit for the development and production at Alamidos Heights in Long Beach.
He was clever enough to sell the leases to his father's
company while retaining a participating interest. By 1928
he was pressing George Getty about the need to achieve
greater economies to boost profits. Again George Getty
made a business transaction with his son, but this time on
a favourable basis. He sold him 33 per cent of George F.
Getty Inc., which held the California properties, for $1
million. Of the $1 million, only $250,000 was paid in
cash; the rest was in the form of promissory notes.

In any normal relationship between the founder of a
company and his only son, a share of the family business
might be offered for love. All the same, this $1 million
investment seems to have been a far better long-term
plan than any of Getty's marriages. At the time that
Getty was buying a share of the family operation, a
prominent California geologist was telling Charles
Wrightsman, an independent oilman, that the Getty
properties were worth $27 million. So Getty picked up a

bargain for a small cash-down payment, one of his favourite tricks.

In 1929 J. Paul Getty had to decide whether he should attend his parents' golden wedding anniversary on 30 October or examine at close hand the crash of the stock market in New York. He chose crucial drama on Wall Street, where he 'sensed that the stock market's collapse and its ramifications would set the course of the nation's overall economy for several years to come . . . I watched the debacle at first hand and talked to brokers and bankers, businessmen and financiers, investors and speculators. What I heard and learned during the fortnight I remained in New York would prove invaluable to me in the years that followed,' he wrote in *My Life and Fortunes*. 'I learned much about the stock market, stocks, investment and the perils and pitfalls of speculation . . . I realized that I was an eyewitness to the violent death of an era.' The trip to Wall Street proved to be a turning point in Getty's practical financial education.

By this time Fini had returned to Germany to have his second child. She was unhappy left alone in his mother's house; she did not speak English well and, only nineteen years old, wanted to return to the protection of her parents. Getty too sailed for Europe. His second son, Jean Ronald, was born on 19 December 1929 in Berlin. In February 1930, however, with Fini back in Karlsruhe with her family, Getty pursued twenty-three-year-old Hildegard Kuhn, a pretty office clerk whom he met in a Berlin dancehall. At first Hildegard did not immediately respond to the advances of this 'elegant' man and refused to go out with him, but Getty persisted with his customary tenacity. She began seeing him socially and had a continuing relationship with him over the next forty-six years. He was to leave her $100 a month when he died in memory of their friendship.

Nevertheless, Getty attempted a brief reunion with Fini in Montreux, Switzerland, one of his favourite mountain resorts. On 22 April he received the news that his father had suffered another stroke and was not likely to live very long. It took him nine days by train and boat to reach George Getty's bedside in Los Angeles, where he maintained a vigil outside his father's room.

George Getty had refused any conventional medical treatment and had been under the care of Christian Science practitioners until his lawyer found him propped up in bed, without proper sanitation, and Sarah in such a state of shock that she could only communicate in writing. Dr George K. Dazey was called in, but neither he nor Christian Science could save George Getty. On 31 May 1930 one of the pioneers of the Oklahoma oilfields and subsequently one of the richest men in Los Angeles died.

5

Takeover Artist

When I'm thinking about oil, I'm not thinking about girls.
J. Paul Getty

His father's death, Getty was to write in his autobiography many years later, 'was the heaviest blow, the greatest loss, I had suffered in my life.' It was also an enormous personal shock to him, for George Getty left the bulk of his estimated fortune of $10 million to Sarah Getty. In addition, he had left control of his estate, and therefore of the Getty oil interests, in the hands of executors from the Security-First National Bank of Los Angeles and H. Paul Grimm, a conservative associate of George Getty. The thirty-eight-year-old Paul Getty was left $500,000, less $250,000 which he still owed his father for the purchase of 33 per cent of George F. Getty Inc. in 1928. Getty's six-year-old son by Jeanette, George II, was bequeathed $300,000.

Getty's accountant at the time, Walter Scott, told Ralph Hewins, Getty's first biographer, 'When his father died, Paul was swindled and hurt, and has since built up a protective armor.' Getty complained in his privately published *History of the Getty Oil Business*: 'I continued the act of management of the company, but was greatly handicapped after May 31, 1930, because I was not an executor of father's estate and did not inherit his controlling stock in the company.'

On 3 July 1930, Getty was elected president of George F. Getty Inc. but, as he explained later in his autobiography, 'final authority rested elsewhere, with the majority stockholder, who, of course, was my mother, counselled and supported by the executors.'

This was to prove a major obstacle to Getty's growing ambition to expand the family oil enterprise into a major integrated oil giant which refined, transported and sold gasoline as well as producing crude oil. He no longer wanted to be at the mercy of the major oil companies which could squeeze a poor independent producer by blocking access to storage capacity, to pipelines and to markets.

Getty saw the depression as a great opportunity to create an integrated oil giant. While most other investors and businessmen were cutting back and running for shelter, Getty had the vision to go against the grain of economic disaster. It was perhaps his greatest stroke of courage and genius. 'I was convinced the nation's economy was essentially sound,' he wrote in *Playboy* many years later, 'that though it might sag lower in the near future, it would eventually bounce back, healthier than ever. I thought it was the time to buy – not sell.' Most observers, frightened by the sudden decline in the price of oil from $1.19 a barrel to 65 cents, thought Getty was making a fatal mistake.

This economic opportunity meant a change in Getty's basic philosophy as a businessman. During the 1920s he made his money in the oil business by drilling wells and producing oil in towns along the California shoreline – Santa Fe Springs, Signal Hill, Athens and Huntington Beach. At the time, the stock prices of oil companies were above book value and overpriced, Getty felt. It was cheaper during the 1920s to buy leases and drill for oil.

In the 1930s, after the crash of the stock market, Getty realized that he could pick up shares of oil companies at a fraction of their value in terms of the oil in the ground. He 'considered that it was foolish to buy oil properties with 100 cent dollars when you could buy them indirectly with maybe fifty cent dollars.' This concept was a basic motivation in Getty's empirebuilding for the rest of his life.

However, in 1930, Sarah Getty was opposed to buying common stock. Nevertheless, in September Getty convinced the directors of George F. Getty Inc. to begin buying shares of two California oil producers, Mexican Seaboard and Pacific Western, which owned valuable acreage in an important oilfield called Kettleman Hills. He purchased $3 million of Pacific Western and Mexican Seaboard stock using a margin account at E. F Hutton and Company as well as a $2.5 million loan at Security-First National Bank. He then left for Europe on an extended trip only to see the stock market suddenly weaken, causing a $1 million paper loss in the oil shares he had just purchased. The bank forced the Gettys to sell the Mexican Seaboard shares to pay off part of the bank loan. He told a biographer many years later that 'the Directors, oilmen and not speculators, were resentful. Mother was vindicated. I realized the atmosphere was decidedly chilly.' The $1 million paper loss must have shaken even Getty. Earlier, Getty's overconfidence in the stock market had produced a loss of $56,677.49 when his private portfolio plummeted in the late 1920s.

This setback was another challenge to his energy, nerve and stamina. Getty said later, 'My stock purchases were financed by every dollar I possessed and every cent of credit I could obtain. Had I lost the campaign (and I was defeated in several preliminary skirmishes and came within hair-breadths of total failure on several occasions), I would have been left personally penniless and very much in debt.'

His first goal was to use George F. Getty Inc. to obtain working control of Pacific Western, one of the ten largest oil-producing companies in California. Pacific Western was a bargain, Getty realized, since its shares had fallen from $17 when he first began acquiring them to a rock bottom price of $3. For one thing, Pacific Western's book value was $15 a share. For another, its oil reserves, located in the California oilfields Getty knew so intimately, were worth far more than $15 a share. By the time Getty finally obtained control of Pacific Western in late 1931, he had used up his own personal credit. Characteristically, he took advantage of the nation's economic disaster by firing every single employee and hiring them back at a smaller salary. His principle of management was 'a lower overhead creates larger profits'. David Staples, then a young Canadian lawyer, had his salary reduced from $400 to $300 a month. There were 12 million Americans on the breadline and jobs were hard to come by.

Next on Getty's agenda was the takeover of an oil company far larger than Pacific Western. Tide Water Associated Oil Company was the ninth largest oil company in the USA, with $200 million in assets and 1233 service stations. Even in the depths of 1932 its net profits were nearly $5 million, almost equivalent to the entire revenues of George F. Getty Inc. At the rock bottom of the bear market, Tide Water shares had fallen some 90 per cent in value and were selling for less than $3 a share. Getty could not help but notice that all of Tide Water, some 5,790,000 shares of common stock, were worth only $14.5 million.

Getty wanted Tide Water because it needed crude oil for its refineries in San Francisco and Bayonne, New Jersey. Getty planned to supply the crude oil produced by George F. Getty Inc. and Pacific Western to Tide Water, which would assure him of downstream markets.

But Getty needed money to go after Tide Water. Thus,

in January 1932, he sold George F. Getty Inc.'s valuable acreage in the Kettleman Hills oilfield located in the San Joaquin Valley for $4.5 million cash to Shell Oil. He boasted to his biographer Ralph Hewins, 'I was the only man in America with big money in cash at the time.' However, this claim ignored wealthy men like Bernard Baruch, who wisely liquidated some holdings before the market collapsed, and a clique of short sellers like Joseph P. Kennedy who pocketed fortunes from the debacle.

Still, Getty's $4.5 million was enough to begin his campaign and he chose an opportune moment in financial history to do so. On 12 March the Swedish match king, Ivar Kreuger, had committed suicide in a Paris hotel, and two days later George Eastman of Eastman Kodak took his own life in Rochester. Anxious officials of the Stock Exchange kept the news off the trading floor until the stock market closed, but next morning the shares of Eastman Kodak, one of the nation's blue-chip companies, took a dive, carrying several other household names with them.

A modest order to buy Tide Water Associated shares, entered on that day of chaotic trading on the Stock Exchange floor by the Los Angeles office of E. F. Hutton on behalf of an unknown investor, went unnoticed. During the trading session of 15 March 1932 Getty bought his first 1200 shares of Tide Water at $2.50 a share, a fraction above the lowest point they would ever reach, and he went on buying. By the end of the month he had picked up 15,100 shares at a total cost of $37,232.50 without giving any indication to the top executives of Tide Water that they would have to contend with a tough, aggressive interloper. 'Buy when everyone else is selling, and hold on until everyone else is buying,' was Getty's credo, a simple formula few others had the patience or the nerve to follow.

As if this exercise were not enough, he was trying at the same time to expand the Getty oil interest into the Middle East. That he tried at all, when his resources at home were stretched so thin, shows his commercial audacity, but the legend that has grown up around this venture is revealing. He always claimed that the government of Mesopotamia (now Iraq) offered him a concession in 1932 for the proverbial song. To his eternal regret, he turned it down. 'My decision was a classic boner, one that I would rue and regret in the years that followed. I had allowed a fantastically valuable concession to slip out of my hands even though it was being offered at a comparatively negligible price,' he wrote in *My Life and Fortunes*.

That is not, however, the way either the Iraqi Government or the US State Department remembers it. In early 1930 Iraq wanted to increase its oil exports and thus its revenues, and was prepared to offer a concession in an area west of the River Tigris and north of the 33rd parallel. When Getty heard about this later in the year, he started negotiations with the Iraqi Government in Baghdad through an apparently well-connected local agent, Albert Kasperkhan, who had once worked at the Ministry of Finance. There was tough opposition including the Iraq Petroleum Company. This has since been nationalized, but it was then a consortium of Anglo-Persian Oil (now British Petroleum), Royal Dutch/Shell, Compagnie Française des Petroles, Gulf Oil and Standard Oil of New Jersey, not to mention Calouste Gulbenkian, who was known as Mr Five Per Cent because he made his fortune by taking 5 per cent of the largest oil concession in the Middle East. Getty was also bidding against the British Oil Development Company, a group of independents with financial backing from Britain, France, Holland, Italy and Germany, and sponsored by the Italian and German governments, which promised to take some of the oil.

Getty had the advantage of support from the US State Department, which was hoping to establish an 'open door' policy for US oilmen in the Middle East. The acting Secretary of State in Washington, William R. Castle Jr, instructed Alexander Sloan, the US consul in Baghdad, to 'render Mr Getty appropriate assistance'. Eventually Getty met Iraq's Prime Minister, Nuri Pasha, who was initially cool towards the American oilman. Their meeting took place somewhere in Europe.

Kasperkhan, an Iraqi Christian who claimed that his family had connections with King Faisal and with Nuri Pasha, told Alexander Sloan in Baghdad that Getty 'was soon able to convince the Prime Minister that he was powerful in the United States not only financially but politically, and after that he was listened to.' Sloan passed this information on to the Secretary of State. According to Sloan in a confidential memorandum dated 29 December 1931, Getty told the Iraqi Prime Minister that he had men and equipment to start drilling within six months; that he would build a pipeline from the producing fields to the Tigris; that he would transport the oil on river tankers or barges to Basra and then by ocean-going barges to markets in India, Japan and China, which, he claimed, would take any amount of Iraqi oil he could lay his hands on. He also offered the Iraqis higher royalties on production than any other company.

These claims were inflated and, at that stage in his career, unfulfillable, but he did not stop there. Again according to Sloan, he told the Iraqis that he was 'on friendly terms with government officials from President Hoover down' and that he could guarantee American political aid to Iraq. Not surprisingly, the Iraqi Government checked up on George F. Getty Inc. and took up its financial references. They found that 'the company was worth about $30 million . . . and could not invest in Iraq in amounts sufficient to export oil in commercial quantities in the near future.' In fact, in 1931, George F. Getty Inc.'s gross income was $2.3 million, while its profits were $148,588.70. The concession went to the British Oil Development Company, later to be absorbed by IPC.

Getty seems to have picked the wrong negotiator in Albert Kasperkhan. The British adviser to Iraq's Ministry of Economics and Commerce, H. H. Wheatley, described him as a man 'who had no influence and who is considered to be not the type which an oil company really desirous of operating in Iraq would select as their representative.' Nor was the government 'impressed with the George F. Getty Oil Company's attitude prior to submitting the tender.' Sloan reported to the Secretary of State that Getty's offer was 'made in such general terms as to offer very little chance for even beginning negotiations.' Getty may have been able to mislead the wildcatters of Oklahoma over the Nancy Taylor lease, but he could not fool the government of Iraq or overcome the competition of far better capitalized oil companies.

The British Oil Development Company was in a far better position to market Iraq's oil production in Western Europe. At the time Iraq was counting on Western Europe's help in obtaining recognition from the League of Nations. Moreover, the well-capitalized European group was far more able than Getty to pay the mandatory $1 million minimum annual payment for the concession. Getty was not yet a Gulbenkian who could carve out a 5 per cent interest in Middle Eastern exploration and production. He did not have the financial and political clout to obtain oil deposits which were to become commercially viable after the Second World War. By then

Getty would be back in the Middle East, when it made a great deal more economic sense for him personally. In the meantime he had more pressing matters to deal with back home in Los Angeles, where his personal life was bizarre beyond belief.

6
Double Bigamy

My wives married me; I didn't marry them.

J. Paul Getty

It is scarcely credible, given the hectic pace of his business life in the early 1930s, that Getty had either the time or the energy to devote to a private life which was, if anything, even more feverish. He expressed his own attitude to women in no uncertain terms. 'While I did not purposely start romances with other women, I did have the philosophy that many men have but which is anathema to their wives. I had many good women friends and I felt that just because I was married was not a reason for me to deliberately avoid these women,' he wrote in *As I See It*, his posthumously published autobiography.

Part of the reason for his womanizing in later life was the failure of his early romantic attachements. In 1973 he told an interviewer that his one great disappointment was 'as a young man being attracted to many beautiful girls, who in turn were attracted to other young men.'

His marriage to Fini, who was still in Karlsruhe, was falling apart and in November 1930 he left for Berlin to begin negotiations for divorce. But ever since he had met and dated the fourteen-year-old schoolgirl in Hollywood eight years earlier, he must have been obsessed by thoughts of Ann Rork. He had telephoned her from time to time asking for a date, but her parents refused to allow it. It was not until after they had been impoverished by the Wall Street crash of 1929 that they allowed their daughter to see the multimillionaire tycoon. Ann was only twenty-two, small and well covered with baby fat, the result, perhaps, of all the milk and crackers she had consumed at boarding school. Her father gave her a part in *A Blonde Saint*, which starred Gilbert Roland, one of Hollywood's great matinee idols. Getty's courtship of her was peculiar, to put it mildly. 'He was like a great big lumbering bear. He was really very prudish and I knew so little about him because my early dates with him were brief,' Ann recalled many years later.

The thirty-eight-year-old oilman, already a married man three times, took Ann to Berlin in 1931 and set her up in an apartment in the same building as his own. She regarded him as her saviour from poverty and, although the relationship began as a matter of economic necessity, she quickly came to love him. 'I thought he was God,' she said, 'his knowledge was awesome.' Getty was her first lover and 'a very considerate one at that. I was introduced properly. And I certainly hope that I pleased him.'

The divorce from Fini, Getty's third wife, was proving not quite so easy as his first two. For one thing, he was forced to withdraw his application for a German divorce when Fini produced a signed report from a private investigator, hired to follow Getty, about his affair with another German woman, whom he had taken on a trip to Dresden.

In August 1931 Getty and Ann returned to New York from Berlin and moved into the Plaza Hotel on Fifth Avenue. And there, in the following months, Getty made a matter-of-fact proposal of marriage. 'I want to marry you. Do you want to marry me?'

Ann answered, 'I do.'

'Then we'll marry,' said Getty. 'We don't need any third person. We don't need any licence or ceremony.' And, indeed, the only record of a marriage at this stage is Getty's claim on a passport application that they were married on 5 October 1931. This claim was a clumsy attempt to cover up the out-of-wedlock birth of Eugene Paul, for, early in 1932, Ann became pregnant.

As always with Getty, this marked a turning point in their relationship. He left immediately for Europe. True, he cabled to Ann from Paris to join him there, but when she arrived he went out with another woman on their first night and left Ann alone for ten miserable days in the dark and gloomy apartment which he used as a love nest on the Rue St Didier. According to her divorce papers, the wretched girl swallowed a bottle of iodine in a clumsy attempt to commit suicide, but Getty returned to the apartment in time to rush her to the American Hospital, where she recovered without losing the child.

From Paris they set out on a grand tour of the European capitals. Often he ignored her, choosing instead to visit night clubs and restaurants with other glamorous friends. In Naples, in late August 1932, he insisted that Ann, now eight months pregnant, accompany him on an expedition up the steep, dusty slopes of Mount Vesuvius and into the crater, which she found 'filled with oppressive smoke and gases'.

A few days later, on 7 September 1932, on board a freighter bound for Genoa, Ann gave birth to Getty's third son, Eugene Paul Getty. His birth was legitimized some three months later when his parents married. He was meant to be his father's namesake and later changed his name legally to Jean Paul Getty, or J. Paul Getty Jr.

On arrival in Genoa, Getty went to the American consulate and registered the birth of his son, claiming, as he had on the passport application, that he was married to Ann Rork. Perhaps it seemed to him little more than a white lie, for there is no doubt that he planned to marry her after his divorce to Fini came through. But the lie was later to land him in trouble with the American authorities.

Meanwhile, he left Ann and his newborn son alone in an Italian hospital, while he headed for Paris and his love nest. When she had recovered sufficiently to join him there, he gave her a limited amount of money for meals, refused to hire a nurse or any kind of help for the child, and left almost immediately for the United States. In November 1932 she received a frantic telephone call from him asking her to return immediately to help him straighten out a serious problem with the State Department. He had committed a criminal offence by declaring he had married her. He told her to meet him in Cuernavaca in Mexico.

There, on 1 December 1932, she signed a prenuptial agreement, in Spanish with no English translation, and therefore incomprehensible to her. By her signature, Ann renounced her rights under the community property law of California. If they should ever divorce, he agreed to pay her $275 a month alimony but, as in the case of a similar agreement he had made with Fini, should she commit adultery, the alimony would last for only one year.

Getty married Ann even though he knew that Fini had not signed the divorce papers he obtained in Cuernavaca in August 1932. Apparently this divorce was not valid because sixteen years later, in 1948, Fini returned to Mexico and finally accepted a divorce. She testified in 1984 that 'Paul was not on good terms because I was fighting the divorce, my divorce, and he had applied for divorce in Cuernavaca.' Her father, Otto Helmle, wrote to Getty in October 1934, more than two years after the divorce was supposed to have been final, threatening to visit the American consul to inquire 'whether the lawyer's interpretation that the authorities won't recognize the divorce is correct.' In addition, Ronald Getty has testified that in 1948 he accompanied his mother, Fini, to Mexico on the 'understanding that it had to do with re-doing a divorce . . . they had to go for the divorce.'

Even Getty may have had doubts about the conclusion of one marriage and the beginning of the next, for when Ann later began divorce proceedings against him he tried to claim they were not properly married. Nevertheless, after the Mexican ceremony, on 2 December 1932, he dispatched his young bride to Washington DC where she went to the State Department in order to try to persuade the Assistant Secretary, Mrs Ruth Shipley, not to prosecute Getty for making false statements on his passport application form. Ann must have been convincing, for the State Department dropped the matter.

Getty tried to keep the most embarrassing parts of his chaotic personal life from his mother. Just two weeks after his secret marriage to Ann in Cuernavaca, he told Sarah, 'I wish I were with you. I think of you so often. There's no one can take a mother's place and you have always been such a kind, loving mother to me.' This may have been a softening-up process because, instead of breaking the news about Ann and the birth of Eugene Paul, Getty went on to tell his mother about the divorce from Fini which he claimed to have just got. He explained that he was afraid to come to Los Angeles because Fini might serve legal papers on him in an attempt to get more money.

When Ann returned to Los Angeles, she and her son were lodged in a small, cheap apartment, while Getty went to his mother's house on South Kingsley. For months he did not tell Sarah about Ann or Eugene Paul, although he came and went freely to and from Ann's apartment. 'He didn't want to hear or smell children,' Ann was to comment on this period of their married life. 'He wanted them for his dynasty but he didn't want to deal with them as babies.' But still she could not tear herself away from him. She was, in any case, totally dependent on him financially. On St Patrick's night, 17 March 1933, he made love to her standing up in the kitchen and the result was another pregnancy.

Almost immediately Getty was off again, this time using as an excuse his fear of the earthquakes that momentarily caused turmoil at Long Beach, not far down the coast. Ann recognized that he was 'not really and truly a bold man except at investments'.

He stayed away until the end of 1933, the full nine months of her pregnancy, spending the time in Germany and seeing at first hand the spectacle of Adolf Hitler's rise to power and his ruthlessness in removing all opponents and dissenters. One business executive who fell by the wayside because of his opposition to the Nazis was Otto Helmle, director general of the Badenwerks electricity company and Getty's formidable antagonist in the divorce proceedings. Getty admitted to Fini later that he was not sympathetic to Helmle's fate. Indeed, Fini testified in a 1984 court action, 'Paul told me that he had admired the Nazis very much and he had been in Berlin, spent a lot of time in Berlin, had met many of the top people . . . He even met Hitler and Hitler had given him a necktie, made him a present.'

When Ann's father, Sam Rork, died suddenly in July 1933, Getty arranged for her to move in with her widowed mother to 'save on living expenses'. On 20 December, she gave birth to Getty's fourth son, Gordon. Getty himself was now back in California. He arrived at the hospital three hours after the birth, stayed for a moment to see the baby, commented, 'Uh, he looks like you,' and left.

Ann and the two boys moved into a rented house in
Santa Monica, California, in February 1934. Her husband
often came home late at night, using the age-old excuse
'I've been at the office.' Ann found letters from other
women, was forced to unpack his luggage when he
returned home from trips and found his contraceptives
and other sexual paraphernalia. According to Ann, he
even showed her a one-page legal agreement, drawn up
by Thomas Dockweiler, his legal adviser, which he asked
women to sign before he bedded them. It waived any
claim against him if they became pregnant. He liked to
try to pick up women on street corners or other public
places, to see if they would go to bed with him without
knowing his identity. These anonymous assignations were
to prove to himself that he could be loved for reasons
other than his name and his wealth. Sometimes, when he
and Ann went to the movies, he would arrange for his
old flame, Belene Ashby, to meet them. He liked having
more than one woman admirer around.

Still Ann held on. She was waiting for him to complete
building an eight-room cottage, Cape Cod-style, overlooking the Pacific Ocean, just down the beach from San
Simeon, the gigantic castle that William Randolph Hearst,
the publishing czar, had built for the actress Marian
Davies. The grandeur of Hearst's five interconnecting
colonial buildings, with a total of 110 rooms and fifty-
five bathrooms maintained by thirty-two servants, made
Getty's new home look like a gatekeeper's cottage. He
wanted to emulate Hearst's grandiosity, even compete
with him. In his diary he wrote: 'I guess I'm like Hearst
in that I admire splendor. I like a palatial atmosphere,
noble rooms, long tables, old silver, fine furniture.' In
later years he compared the lengths of their respective
swimming pools and refectory tables. In the 1950s, after

Hearst died, Getty even considered buying San Simeon as his worldwide headquarters, but decided it was located too far from civilization.

While in one sense he took the great newspaper proprietor as a model, Getty was a totally different personality. Hearst was gregarious, with a gargantuan appetite for life, generous to a fault, while Getty was cold, aloof and miserly. Although Hearst did not speak to his wife for twenty-five years, he was the constant, devoted partner of Marion Davies. Not one of Getty's women would be able to say, as Marion said of Hearst, 'Companionship and devoted love – that was our pact.' She admired him because 'he had no idea of money at all. It didn't make any difference to him.' Getty, on the other hand, always brought business papers to the dinner table. 'He was so obsessed with business,' Ann wrote, 'he had little interest in anything else.'

'I suppose that in 1935 I was one-third as rich as Hearst and by 1950 I was twice as rich,' he wrote in his diary. Hearst borrowed $1 million from a mistress and gave her two Boston newspapers as collateral. By contrast, it was always Getty who took collateral from the women who borrowed from him. He made Ann lease the beach house in Santa Monica and pay the Santa Fe Investment Company $200 a month rent out of her monthly allowance of $600.

Still, there were outings on one of her husband's yachts, albeit infrequently. He purchased his yachts from wealthy industrialists and then traded them off for larger ones, the last one being utility magnate Harrison Williams's 260-foot ocean-going vessel. Getty refused to man her adequately because of the cost. There is little evidence he got much enjoyment from his vessels, but there was a political reason for them. Getty was worried about the

Communists, he told Ann, and he explained to her that he needed to know that he could sail away at any time if he was forced to escape from the enemy who might want to take his millions away from him.

A Mother's Trust

There are one hundred men seeking security to one able man who is willing to risk his fortune.

J. Paul Getty

Neither the acquisition of yachts nor of beautiful women was allowed to stand in the way of Getty's driving ambition for power in the oil industry. To build his empire, he needed to achieve several financial coups in succession. First, as a minority shareholder of George F. Getty Inc., he needed to acquire his mother's two thirds ownership in order to obtain total control of the company. Then, once he had 100 per cent of George F. Getty Inc., Getty would be free to borrow additional money in his quest for the much larger Tide Water Associated, in which he had already purchased a small holding.

Sarah Getty was a tough and formidable opponent to Getty's ambitions. She had been a director of the Minnehoma Oil Company since 1906 and a director of George F. Getty Inc. since it was organized in 1924. 'I thought that mother was continuing to interfere with my plan to purchase stock in Tide Water Associated,' Getty wrote in *The History of the Getty Oil Business* in 1938. 'Mother was conscientious in studying the company's affairs, and I often marvelled at the shrewdness of her judgment . . . [But] I was convinced in my own mind that the Getty companies should use all their resources to continue acquiring Tide Water Associated stock.'

The relationship between mother and son was unusual, even bizarre. On the surface Getty told everyone that

Sarah was the finest mother in the world. After all, respect for one's mother was expected from an only son, especially if she was old, lame and deaf. In fact, he could never make the right connection with her. He kept an apartment full of clothes at her home and used her address as his legal residence, although he was supposedly living with Ann in Santa Monica. (He also kept on a bachelor hideaway, described by Dave Staples, one of his employees, as 'a bohemian love nest', up in the hills where he could take other women.) He provided Sarah with plenty of grandchildren – the dynasty that Hearst had told him was essential for continuing the family tradition.

In return Sarah treated him like a little boy, calling him, at forty-one, 'my dearest child'. She worried each time he married that she might be losing him for good. 'Don't take my boy away from me,' she told one of the wives. Perhaps this possessiveness accounts for the fact that he introduced none of his wives to her before getting married. For her part, she was shocked and upset at his four marriages and three divorces inside a decade. They demonstrated an instability and immorality she found impossible to understand.

In 1933, Getty was in the prime of life, full of furious energy, 'a wild egg who didn't give a damn for anyone', in the words of Dave Staples. Sarah, on the other hand, was eighty years old, grossly overweight, dressed constantly in black Victorian clothing and crippled with rheumatism. She was also completely deaf, and visitors had either to shout through a black amplifying box or resort to written notes to communicate with her. To get her out of the mansion and into her touring car required the muscular arms of two men who were by her side day and night. But she still had most, if not all, her wits about her.

During the year 1933 the two stockholders of George F. Getty Inc. – Getty with 33 per cent and his mother with 67 per cent – worked 'at cross purposes'. Friction between mother and son was exacerbated early in that year when George F. Getty Inc. purchased Sarah Getty's personal holding of Minnehoma Oil and Gas for $1,609,000 to pay her husband's death duties of $1.3 million. Getty complained that the purchase price of $2.50 a share was 50 cents per share more than Minnehoma stock was worth. He argued in *The History of the Getty Oil Business* that the transaction which paid his father's estate taxes was unfair to him as a minority stockholder and had cost him $107,000.

Getty also objected vociferously to loans made by George F. Getty Inc. to a separate company, George F. Getty Oil Company, completely owned by Sarah, in order to enable it to acquire leases and drill wells in New Mexico. So acrimonious were the fights on the board of George F. Getty Inc. that Sarah Getty's adviser, Rush M. Blodget, the company's chief lawyer, resigned in 1933, claiming that opposition to her son was 'a heavy drain upon my vitality'. He recommended that Sarah Getty sell out for $1 million in 'good interest-bearing securities'.

The family dispute continued until Christmas Day 1933, when Sarah finally offered to part with her two thirds interest in George F. Getty Inc. for $4.5 million. Getty was to pay his mother, not in cash, but with interest-bearing promissory notes that would give her an income of 3.5 per cent and could be cashed in by her over the next several years. In return, Sarah Getty would give her son a Christmas present of $850,000 to pay off his margin debt to E. F. Hutton and another obligation to George F. Getty Inc. Sarah Getty was a tough negotiator, for she informed her son that the offer would last only until noon on 30 December and 'if not accepted by you in writing on or before that date and hour shall be considered as withdrawn by the undersigned and shall be wholly terminated and at an end.' It hardly sounds like a letter from a loving and generous mother to her only son. The deal went through, but if either thought it was going to resolve their differences, they were mistaken. All Getty had really accomplished was to turn his mother from his largest shareholder into his largest creditor and an interfering one at that. Worse still, now that he owed his mother $4.5 million, the banks refused to lend him sufficient funds to finance his expansion plans for taking over Tide Water. He felt he was mission one of the great investment bargains in history because the stock was moving up all the time.

Throughout 1934 Sarah kept on insisting that the company stop buying Tide Water Associated stock and constantly worried that, if the stock market turned, the value of her notes would be destroyed. It is not surprising that Getty was in a hurry to get rid of her as a creditor so that he could remove the $4.5 million debt from the books of George F. Getty Inc. and obtain a fresh cash infusion to buy more Tide Water. Indeed, he needed to hurry because Standard Oil of New Jersey, the Getty family's archenemy and the largest stockholder in Tide Water Associated, was planning to organize a new company called Mission Corporation to hold the block of Tide Water shares. Getty needed money desperately to buy Mission Corporation shares as a way to continue building his position in Tide Water. 'It was not beyond the means of the Getty company to purchase control, providing that the company's financial structure could be cleaned up,' he testified several years later in a court action. 'It [George F. Getty Inc.] had a very heavy . . . short-term indebtedness and as long as that existed, it was impossible for the company to have any sort of proper credit rating;

so I urged my mother to make a gift of these notes to me.'

He began by trying to convince his mother that she did not really need the annual income of over $140,000 from the interest-bearing notes, because her annual expenses were only $25,000. He also attempted to worry her about the problem of reinvesting the capital represented by the notes which she was scheduled to receive over the next nine years. What he really wanted was a gift of the $4.5 million notes for Christmas 1934; indeed, he asked her for them outright, but 'found her rather unreceptive to the suggestion'. After her rather generous gift of $850,000 the year before he seemed mildly surprised. He worked on her assiduously, often staying at her house rather than returning home to Ann and the two boys in Santa Monica.

And so the duel between these formidable opponents continued for weeks. Sarah probably knew by now that Getty would somehow get his own way in the end. Her main concern was not herself. She had enough to live on for the rest of her days, but if she relinquished control of George F. Getty Inc., she wanted to ensure that her husband's legacy would be passed on to future generations. How to keep the nest egg from being ruined by her son's ambitious plans?

Sarah Getty's solution was to establish a trust that would protect Getty and his children, while still allowing her son to make use of the money to expand the oil company. She would contribute $2.5 million of the notes issued her by George F. Getty Inc. a year earlier. As part of a trust it would be protected 'from any unfortunate financial condition' Paul Getty might get into, and that included bankruptcy.

Sarah also demanded that her son make a cash contribution. Getty resisted having to dig into his own pocket. 'I, of course, was not in favor of making any contribution myself to the trust because I needed every penny I could get my hands on. I was short of money,' he said in a deposition for a lawsuit many years later.

In the end he agreed but, by a typical sleight of hand, he contrived to make his contribution of $1 million without having to use his chequebook. Well, not quite $1 million; in fact, $868,000.

This he did by exchanging some shares he owned in George F. Getty Inc. for a $350,000 note already contributed to the trust by Sarah. He told his mother his shares actually were worth $1,218,420. Therefore, after deducting the $350,000, Getty was able to claim a contribution of $868,000 – only just short of the million his mother had asked for.

In fact, Getty's fourth son, Gordon, was to complain in court many years later that his father had never intended to make a contribution, adding that it was 'vigorously resented, finally acceded to, after considerable disagreement as to its size, as the only alternative to receiving the use of no assets at all.'

Getty became the sole trustee of the trust and was given almost absolute power to arrange any financial transaction involving the Getty oil interests. However, he was prohibited from investing in the common stock of companies other than those in the Getty complex. The trust allowed Getty to invest spare cash only in interest-bearing bonds of the USA, Great Britain, Canada, Norway, Sweden, Denmark and Switzerland. This limitation meant that Getty had to concentrate on protecting the Getty oil operations and could not diversify the trust into any business he liked. Still, there was ample compensation. The income from his contribution, rated at 20 per cent of the whole, was to be set aside for his sons' financial support. But the income from Sarah's contribution, some 80 per cent, would go to him. This protected Sarah's contribution from being taxed at her death. And as Getty's oil interests expanded, the trust began to produce a considerable cash income, helping Getty to enjoy the fruits of his mother's legacy.

Sarah may have been tough-minded, but Getty had got his way without giving his mother or the trust a dollar. While writing *The History of the Getty Oil Business* a few years later, Getty confided to his aide, Howard Jarvis, a publicist from Utah who later became a famous advocate of lower California property taxes, 'I just fleeced my mother.' Jarvis, a devout Mormon who respected his own mother more than anyone else in the world, was quite shocked and even briefly considered quitting Getty's employ. He told an interviewer years later, 'I thought he should not have pulled a sharp deal on his own mother. I thought that was unethical.'

Neither mother nor son could have imagined to what extent the Sarah Getty Trust would become the vehicle for one of America's great fortunes. The trust began with a value of $3.368 million on 31 December 1934. Its goals were modest. Sarah wanted to make absolutely sure that her son took care of his primary financial obligations – support for his wife Ann and his children from three marriages, George II, who was ten, Ronald, who was four, Paul Jr, two, and Gordon, one. The amounts settled on the boys in 1934 were sufficient for their needs but not stupendous. After Ann received the first 10 per cent of the trust's income, the next $21,000 was divided between three of the sons – Paul Jr $9000, Gordon $9000 and Ronald $3000. George was left out of this first distribution because of his grandfather's bequest of $300,000 in a trust fund in 1930. If the Sarah Getty Trust earned more than $21,000 in any given year, then Paul Jr and Gordon would share equally in the excess. Ronald was excluded from this potential dispensation. The unequal distribution

of the trust's income was always a sore point between
Getty, Ronald and Ronald's mother, Fini. It seems to
have been Getty's way of taking revenge on Fini and
her father, Otto Helmle, who had attempted to obtain
increased alimony payments for his daughter. On 20
November 1934, just before the trust was created, Getty
warned Helmle, 'You cannot expect if Ronnie's mother
adopts a course of conduct which brings in worry, anxiety
and trouble that I will feel disposed to leave Ronnie a
large sum of money.' In fact, many years later, Ronald
believed that his unequal treatment was because 'my
mother against his wishes had taken me to Germany to
live and also she said that she didn't want to come back
to the United States . . . and that he expected that I
would be inheriting a lot of money from that side.' Getty
later described Helmle as 'a businessman who was most
certainly my equal'.

To this day Ronald has received only $3000 a year in
income from the trust. As the Sarah Getty Trust multiplied, it made George II, while he was alive, Paul Jr and
Gordon independently wealthy. In 1934, however, no
one dreamed that the income to be divided up among
Getty's offspring in 1985 would be more than $1 million a
day.

8

The Tide Water Mission

The businessman who moves counter to the tide of prevailing opinion must expect to be obstructed, derided and damned.

J. Paul Getty

Now that he was more or less in full control of the family finances, Getty's chief opponent was no longer his mother but the giant Standard Oil Company, who effectively controlled Tide Water Associated. The attempts to take over one of the nation's hundred largest companies was a formidable and ambitious undertaking for a forty-two-year-old independent oilman. He needed to combine the flair of a gambler with methodical planning, but he possessed both qualities in abundance and combined them with a concentration and thoroughness amounting to an obsession which amazed his business associates.

Throughout the 1930s, while the USA was struggling to recover from the worst economic setback in history, Getty was operating at full stretch in the stock market, amassing shares, masterminding intricate internal transactions within his own empire, unseating the directors of companies he had in his sights. He was one of the first modern corporate takeover strategists to use the stock market in order to gain control of other companies, but in one significant sense he was a different animal from today's empirebuilders. He would never have dreamed of borrowing a billion dollars from the banks in pursuit of his ends, or of paying a 50 per cent premium over and above stock market prices to win his battle. What he always wanted was a bargain, an investment to be acquired below book value, in a company with assets, above or below ground, worth far more than he paid for them. Not that he could have borrowed the money in 1935 even if he had wanted to. In the shaky financial climate of the time no bank was going to lend him millions of dollars to take over another company with a contested bid. Moreover, the terms of the Sarah Getty Trust prevented him from borrowing on its assets.

Apart from that restriction, however, as sole trustee of the Sarah Getty Trust, Getty was able to wheel and deal in almost any way that would benefit his quest for Tide Water. So, in the spring of 1935, he exchanged the notes Sarah Getty donated to the trust for shares in several Getty oil companies and an option on 300,000 shares of Tide Water which were owned by Pacific Western. This transaction was the first of many steps by which the Sarah Getty Trust became the majority shareholder in the Getty oil interests and built its stake in Tide Water. Later, Getty merged several different Getty oil holdings into George F. Getty Inc. He was making good on a verbal promise to his mother in late 1934 to 'consolidate the companies and to turn over the majority of the stock to the Trust,' he testified several years later. By the end of 1936, in fact, the Sarah Getty Trust owned 55 per cent of George F. Getty Inc.

The intricacies of corporate finance, the balance sheets of corporations, were perfect natural resources for Getty's meticulous and scientific collection of data which he used for making business decisions. It was a plodding process, but Dave Staples, by now Getty's chief company lawyer at Pacific Western and later executive vice-president, found him to be amazingly and exhaustively thorough. He seemed to know more about an oil well, a bond indenture or a tax law than anyone else.

Nevertheless the pressure to go after Tide Water was

so intense that Getty took the most unusual step for him of going for advice to a psychiatrist on Park Avenue. Ann went along too, because the doctor thought he could treat Paul better if he knew his wife. According to her, the doctor was of no help to Paul, except to recommend he drink crème de menthe after dinner to relieve the pressure, a somewhat ineffective measure, it might be thought, of easing the stresses of a showdown with the nation's second largest industrial enterprise. For Getty had decided to meet Standard Oil head on in a David and Goliath confrontation. After all, George F. Getty Inc. was a modest operation with revenues of only $1,355,380.15 and profits of $368,797.93 in 1935. In addition, the Getty interests owned two thirds of Pacific Western Oil Company, a small but profitable oil producer in California, which earned $663,592.76 in 1935. By contrast, Tide Water Associated Oil Company had produced sales of petroleum products worth $97,103,028.70 during 1934 and had income from its operations and investments of $8,282,702.08.

Just two weeks before the Sarah Getty Trust was set up, Getty threatened Standard Oil of New Jersey with a lawsuit unless the oil giant gave up control of its Tide Water shares, either by relinquishing its voting rights, distributing the stock to its own shareholders or selling the shares in the open market. He charged that the anti-trust decision breaking up the Standard Oil monopoly had made their possession of Tide Water shares illegal. Standard Oil's response to this move was to transfer its 14 per cent interest in Tide Water to Mission Corporation, its newly formed company far away in Utah. Mission was a holding company, formed simply to hold the shares of other companies.

To complicate the matter even further and thus to give the move more teeth, Standard Oil planned to distribute the Mission shares to its own shareholders, including relatives of the founding father, John D. Rockefeller. These included his son, John D. Rockefeller Jr, who had 10 per cent of Standard and so was in line to receive 10 per cent of Mission Corporation.

Getty decided to go after Mission shares as an indirect way to control Tide Water. He got his first chance on New Year's Day 1935, while he was paying his first and only visit to Hearst's castle, San Simeon. While there, he received word that he could buy John D. Rockefeller Jr's rights to his Mission shares.

The New Deal of Franklin D. Roosevelt was planning to scrutinize the power and fortunes of the nation's mightiest families. Rockefeller, who devoted his life to the Rockefeller family charitable causes, was determined to avoid the glare of bad publicity as Standard Oil's largest shareholder. He wanted to avoid the requirements of the new Securities and Exchange Commission, set up in 1933, which would force him to disclose his Standard Oil shareholding. Moreover, he wanted to prevent this interest from being subject to the new inheritance tax which was being proposed in Congress and could have reduced his net worth by 70 per cent. By the end of 1934 Rockefeller shrewdly began to parcel out his Standard Oil shares to his children, including Nelson and David Rockefeller, and grandchildren in private trusts where the shares would not be subject to tax. All this meant that he had no interest in holding 10 per cent of Mission.

Getty's friend, John Jay Hopkins, an investment analyst in New York who was later to found the General Dynamics Corporation, contacted Rockefeller and bought his shares of Mission fo $1.2 million. Rockefeller knew that Getty was the real purchaser, but he had no idea that Getty was on a collision course with Standard Oil of New Jersey for control of Tide Water. As Getty put it in *My*

Life and Fortunes, his roustabout days in the Oklahoma oilfields had taught him that 'the bigger the bully, the better the brawl was likely to be, that no matter how tough the opponent who backed you into a corner, there was always a good chance of outslugging, outboxing or outlasting him in the infighting.'

Having availed himself of the opportunity to acquire Rockefeller's 10 per cent of Mission, Getty quickly and forcefully went after more stock. It took him all of 1935 and 1936 to acquire some 641,000 shares or 40 per cent of Mission. It was still not a majority, but he owned the single largest block of shares and was able to watch over Mission's interest in Tide Water. Faced with this threat, Tide Water changed its directors' term of office from one year to three years. Then it added ten new directors, packing the board with its establishment friends like Henry W. de Forest, chairman of the Southern Pacific Railroad, and Adolphe Boissevain, a Dutch financier who represented a large block of Tide Water stock. This interest, registered anonymously in a Dutch government office, grew mysteriously to equal the 8 per cent interest which Getty now owned. Tide Water also issued both common and preferred shares in order to reduce Getty's voting power.

Nevertheless, he had every reason for satisfaction and pleasure at the highly profitable results of his bargain hunting so far. The battle for control of Tide Water was having a beneficial impact on the price level of its shares. They more than doubled in value from $9⅝ in the spring of 1935 to $20 in 1936, giving the Getty interests a large profit, especially as he had bought his first Tide Water shares for as little as $2.50 each in the spring of 1932. Similarly, the shares in Mission which he had purchased from John D. Rockefeller Jr for a mere $10⅛ on New Year's Day 1935 had almost tripled in value to a price of

$29½ by the end of 1936. One reason for this impressive performance was Mission's ownership of a 55 per cent interest in another oil company, Skelly Oil, the midcontinental producer which had once fallen into dangerous financial straits and sought shelter under the Standard Oil umbrella. Knowing Getty's passion for thorough investigation, it is hard to believe his claim that he was totally unaware that control of Mission would carry the extra bonus of obtaining Skelly Oil. The Skelly shares were on Mission's books at a price of $6.26 each, and by mid-1937 they stood at an impressive $48. After three years of disheartening losses, between 1931 and 1934, Skelly's operation had miraculously turned around and begun to make money. By 1937 Skelly Oil had revenues of $41,500,000 and showed profits of almost $6,500,000, the third highest in its history. This success was largely due to the innovation of gasoline with improved octane quality that minimized the 'knocks' or detonations in the motors, but also to Bill Skelly's nose for oil which had added valuable oil reserves in Oklahoma, Texas and Kansas.

Every time the price of Tide Water or Skelly rose in the stock market, it helped push up the price of Mission shares. This enabled Pacific Western to borrow money using the value of Tide Water and Mission stock as collateral. This money, in turn, could then be used to buy more Mission shares.

The Tide Water and Mission management now thought up another scheme to prevent Getty from getting control. They would simply liquidate Mission Corporation, distributing the shares of Tide Water and Skelly to Mission shareowners. True, this step would give Getty 40 per cent of the Tide Water stock, more than he already had, but this was insufficient to give him total control.

At this point Getty was rescued by a shrewd, tough New York lawyer named David Hecht, who was to become his right hand in every financial and legal machination for the next two decades. Hecht was the antithesis of his client in most ways. He was born poor, the son of a Polish Jewish immigrant. Short, swarthy, with a hook nose and an infectious charm, he had moved through the Townsend Harris High School in Manhattan, spent his college years at New York University, and finished up at Columbia Law School, where he was editor of the *Law Review*. Getty's hiring of Hecht was the classic manoeuvre of a wily fox. He noticed that his New York lawyer, J. Arthur Leve, always consulted a young associate, Hecht, when he needed an answer to one of Getty's problems. One day, instead of consulting Leve, Getty simply called Hecht and apparently asked him to become his lawyer.

It was Hecht who now saved the day. He flew to Reno, Nevada, where Mission was incorporated, and proceeded to block the dissolution. By this time, Getty owned 45 per cent of Mission, and he was able to place a coterie of his associates including Hecht, Dockweiler, his Los Angeles lawyer, Emil Kluth, the geologist, and his friend John Jay Hopkins on Mission's board. What is more, he made Bill Skelly president of Mission and contracted to let the Tulsa-based oilman operate the Getty Oklahoma properties. The contrast between the two men could hardly have been greater. The rough-hewn, bombastic, pugilistic wildcatter, son of a muleskinner who had emigrated from Belfast before the Civil War, was always at loggerheads with a man fourteen years his junior, a smoother, more educated man, son of a wealthy father, with the sophisticated veneer of the big city. As Bill Skelly's official biographer put it, 'The personalities of the two men . . . were not, superficially, the best suited for placid compatibility. Each of them, in his own way, was a little larger than life.'

Getty's control of Mission's board gave him the voting

power which went with Mission's 17 per cent of Tide Water stock. Added to Pacific Western's 8 per cent holdings in Tide Water, Getty now had his hands on over 25 per cent of the vote on any Tide Water matter. Yet the Tide Water group would not give in to its largest shareholder. They listened to his complaints and maintained a civil personal relationship, but he was still an outsider to the Wall Street establishment that influenced Tide Water matters.

'Paul had a lot of points against him,' says David Staples. 'He wasn't in the fraternity. If he had been in the fraternity he could control with 25 per cent.'

Even though he now controlled Mission, it was to be many years more before Getty could dominate the board of directors of Tide Water. Rather, the Tide Water establishment fought Getty at every step. Tide Water acquired other companies to dilute Getty's interest, a technique that made Tide Water an even larger and more difficult entity for Getty to capture; and it used its own employees to organize proxy fights to defeat Getty's attempts at control. In short, this series of takeover battles was a precursor to the modern mania of hostile takeovers involving farflung companies in many industries, but back in the 1930s Getty was one of the few lone wolves trying in an unfriendly way to throw out the incumbent management of a major oil company.

In 1937 Getty began to use the stream of dividends accruing from Tide Water and Skelly Oil to Mission to buy more shares in Pacific Western and Mission. Every time he bought more Pacific Western at an average price of $13.34 Getty was acquiring valuable oil leases in California and the midcontinent area. 'Paul was obsessed by the idea he could get oil reserves cheaper by stock purchases than by developing fields,' Staples commented. His purchases of oil shares, together with the recovery of the stock market, was the cornerstone of the rise of the Getty fortune in the mid-1930s. The Dow Jones industrial average had almost doubled since 1932 and automobile manufacturers, Getty's major market, were looking for their best year since 1929. During the Depression, in fact, gasoline consumption declined less than any other industrial commodity. The demand for gasoline grew in 1934, 1935 and 1936 because of increasing motor vehicle registration and helped to swell the profits of Tide Water and Skelly. It was a powerful argument for owning shares of both those oil companies.

The Getty family fortune had multiplied many times. For one thing, George F. Getty Inc., which Getty owned lock, stock and barrel, was now worth $27,353,977; the rise in the price of Pacific Western, Tide Water and Minnehoma shares had seen to that. (George F. Getty Inc.'s own oil and gas leases were worth a mere $6 million.) For another, the Sarah Getty Trust, which had begun life on 31 December 1934 with $3,368,000 in its coffers, had multiplied in value some six times in four years and now had a book value of $18,573,489. Whatever the eighty-five-year-old Sarah thought about her son's personal life, she must have been impressed by this spectacular performance.

Indeed, it was so spectacular that Getty wished to immortalize it. This saga of oil wells, stock transactions and financial dealings was to be his ticket to fame and glory. He was soon to start work on *The History of the Getty Oil Business*, which was in reality little more than the first volume of an autobiography (two more would follow). In it, he underlined his own role in building the Getty fortune despite his father's mistakes and his mother's obstruction; it was intended to prove how worthwhile his activities had been, at least in terms of dollars, which was the only measuring rod he trusted.

'One in ten makes $50,000, one in a thousand crosses a million, few pass $10 million,' he wrote in the preface to the expensively produced and privately published volume. 'Many are called but few are chosen.' Getty had good cause to regard himself as one of the few.

The Bargain Hunter

I buy when other people are selling.

J. Paul Getty

A fourth broken marriage does not seem to have worried Getty unduly or to have interfered with the rapid expansion of his fortune in the mid-1930s. What does seem to have upset him was the appalling publicity which attended the proceedings when Ann Getty sued for divorce in the Los Angeles courts. His heartless treatment of her now became daily headlines in the Los Angeles newspapers. Ann testified that he had told her she 'was just another woman to him, of whom he had many'. Long after their marriage he was still providing for Allene's sister, Belene, not to mention other women in Germany. And he threatened to throw Ann out of the beach house in Santa Monica where she lived.

Getty reacted to the publicity in the same way that he had reacted to the publicity surrounding the Elsie Eckstrom affair and the sordid divorce from his first wife, Jeanette. He fled. He sought refuge in New York, renting an apartment in Manhattan's most prestigious area, 1 Sutton Place, overlooking the East River. (It was pure coincidence that, twenty-three years later, he was to buy Sutton Place in England from the Duke of Sutherland.)

The apartment, a penthouse owned by a wealthy New York society hostess, Mrs Frederick Guest, was filled with exquisite antique furniture. Here he planned to entertain the cream of the city's social set. And here he decided to shed the shame of the divorce proceedings by becoming a man of culture, developing a love of fine art and indulging his newly developed taste for oil paintings and antiques. Was it pure chance that this decision happened to coincide with a slump in the art market? 'Prices,' he noted in his diary, 'are the lowest of the century.' He calculated that tapestries were selling for 25 per cent of their pre-Depression prices, furniture for as little as 10 to 20 per cent of their prices only two decades earlier, oil paintings at half their value.

There is more than one way of launching out as a major collector. Most of the earlier generation of American millionaires had done so by using the services of the famous, some would say notorious, dealer Joseph Duveen. Getty loved to tell the story of the time Frank Donahue, head of Woolworth's, asked Duveen to send over the finest carpet he had and a beautiful sixteenth-century oriental carpet duly arrived. 'Hell!' exclaimed Donahue when he saw it. 'This thing's been used!' Samuel Kress had handed Duveen $20 million for a collection of old masterpieces, and William Frick gave him $7 million in a single year and a free hand to furnish his fine Fifth Avenue mansion.

Such open-handedness was not Getty's way. He wanted bargains and, with typical thoroughness and planning, he set about mastering the subject which was to dominate the rest of his private life. For a year he hired art experts, starting with Mitchell Samuels, one of New York's most respected and trusted dealers, whose strong suit was furniture and tapestries, to give him private tuition in the subject. He learned all he could discover about the history of furniture, the careers of the craftsmen and the materials they used, the techniques of artists. At the end of the year he was probably as well informed as his tutors.

For a start he decided to concentrate specifically on

those works of art which are overlooked by most collectors seeking paintings – busts of Greek and Roman heroes, and the decorative arts of eighteenth-century France made with impeccable craftsmanship for the Bourbon kings and their queens or mistresses. For some reason he called himself 'Mr G. Paul of California' in the saleroom, or so he was described by the magazine *Art News*. He pulled off a major coup on 22 June 1938 when he acquired for $50,000 a magnificent carpet that had belonged to Louis XIV, a side table with sublime porcelain plaques once possessed by Louis XV, and sixteen chairs, settees, screens and other objects. He noted in his diary that the bidding was 'sluggish, at times becalmed'.

The paintings belonging to Princess Beatrix de Bourbon-Massimo, a descendant of the Comte de Chambord, went on sale at Sotheby's a month later. For a mere $200 Getty acquired a painting of the holy family, known as *The Madonna of the Loretto* and attributed to the school of the sixteenth-century master Raphael. Gerald Brockhurst, the British artist, believed that Raphael had painted part of the picture, and much later in his life Getty maintained that this $200 investment was the equivalent of another Tide Water.

Over the next several months, as war scares mounted, Getty assiduously added to his collection. Gainsborough's portrait of James A. Christie, the auctioneer, was purchased for $37,500, the same price as it had sold for ten years earlier. A jewel-like Ardabil carpet was acquired for only $68,000 personally from Duveen, who was dying from cancer and made a profit of only $11,000 on the carpet which he had bought some twenty years earlier.

It was in November 1938 that Getty made one of his biggest coups. He purchased an early Rembrandt, the portrait of Marten Looten, a wealthy Amsterdam grain merchant, for only $65,000, a third of the price paid for it

in 1928 by a rich Dutch businessman. The portrait held a
special fascination for Getty; he identified with the
sombre merchant who appears to combine a calm dignity
and cool reserve with sensitivity and a feeling of insecurity. This acquisition outraged the Dutch art world,
especially since J. P. Morgan had removed another
important Rembrandt portrait from Holland some years
earlier. This was the crowning point of Getty's prewar
collection which went on display at the World's Fair in
Flushing Meadow in 1939. For $220,000 he had put
together the nucleus of an art collection that would
quickly multiply in value many times over, though perhaps not as fast as his Tide Water shares. Curiously, he
never relates whether Sarah was impressed by or even
curious about his foray into the art-collecting world.

If the pursuit of art was Getty's main interest in New
York, it certainly left him time for other activities, notably
for the pursuit of women. The new object of his attention
was a young auburn-haired night-club singer who was
also high in the social register. Her name was Louise –
Teddy – Lynch; she was vivacious, outgoing, theatrical
and buxom, and she was the right age – twenty-three.
The attractive seems to have been mutual. To Teddy, he
looked like a combination of Leslie Howard and Jean
Gabin, of beauty and the beast. 'He seemed to know
everything,' she enthused to an interviewer, 'especially
about music and art. He was very, very, intriguing.' Most
intriguing of all was his interest in her voice. He thought
she should train for the opera.

They first met at Mon Paris, a Manhattan town house
turned into an elegant French night club, with chandeliers
and red velvet on the walls. From Mon Paris, she moved
to the Stork Club, where the social world, café society
and Hollywood mixed. She would sit at Getty's table,
and Sherman Billingsley, the proprietor, would send over

champagne for them to drink. After Ann had got her divorce, Mrs L. W. Lynch of Greenwich, Connecticut, announced her daughter's engagement to Getty and said the marriage would be in the autumn. However, the marriage did not take place then. For one thing, Getty was preoccupied with several other women and, secondly, he was travelling for almost half the year in Europe.

But another of his preoccupations, as always, was following up investment opportunities at well below their original value. In 1938 he was offered a choice piece of Manhattan real estate on which stood the elegant forty-two-storey Pierre Hotel, on the corner of Fifth Avenue and 61st Street, facing Central Park. Built as a luxury hotel in 1930 just as the Depression hit America, the Pierre had never really taken off and by 1938 was losing money fast. Getty picked it up for $2.35 million, less than 25 per cent of the original cost of construction.

As a member of New York's smart set, Getty needed to attract his playmates and their friends to the Pierre. To make it the in-place for New York's sophisticates, he installed William Rhinelander Stewart, one of New York's most sought-after dinner guests. Wherever Stewart stayed was the place to be seen. The Pierre never looked back.

At the same time that he was establishing the hotel, Getty was trying to obtain a post for himself in the Roosevelt administration. It started with his attempt in 1937 to become an American diplomat, something that he claimed he had always wanted to be. He enlisted the services of June Hamilton Rhodes, an energetic, charming and intelligent lady whose grandmother was a first cousin of Sarah Getty. She had built up her own public relations firm in New York City and during the presidential election of 1932 had worked with Eleanor Roosevelt, encouraging women voters to back Franklin D. Roosevelt. In early 1937 she wrote to F.D.R.'s private secretary, Missy

LeHand, requesting an appointment for Getty with the President to discuss the possibility of a diplomatic post. Her picture of Getty was positively saintly. He wanted 'to devote his life to service and can afford to do so'; his wealth would allow him to 'maintain a diplomatic post properly'; he was 'beautifully educated', the marrying kind, and not a drinker or a gambler. She also mentioned that he had been divorced twice, instead of four times.

There was no invitation from the Oval Office. Instead, Getty saw Forbes Morgan, a member of the wealthy Morgan banking clan, appointed by F.D.R. to screen likely candidates for the top jobs. Morgan thought that Getty was 'the proper type and eligible', but had reservations and passed the buck back to June Hamilton Rhodes: if Getty wanted a diplomatic post, she would have to press the President herself. Perhaps she too was beginning to have doubts for she never pushed the issue and Getty's first attempt to represent his country abroad came to nothing.

But he did not give up easily. In April 1938, a month after Hitler's armies marched into Austria, the US State Department suggested to its friends in Europe that a committee should be set up to aid political refugees from both Austria and Germany. Shortly afterwards, James Roosevelt, the President's son and personal aide, received a letter from the Boston lawyer William Gaston pointing out that the *New York Times* columnist, Arthur Krock, had suggested J. Paul Getty as a suitable candidate for the refugee commission. Gaston took it upon himself to propose Getty as Ambassador to Persia instead, then the largest Middle Eastern oil territory. However, Getty had no wish to go to Persia because 'it is so inaccessible, and without telephone service', and he would not be able to keep in touch with his business interests. The suggestion that he might be asked to join the refugee commission was finally killed by June Hamilton Rhodes herself, who seemed to be rapidly changing her tune about her cousin's suitability. 'This whole Getty business is pretty messy,' she wrote to James Roosevelt, adding that the best thing 'is to forget him and his millions'. Enclosed was a news clipping from one of the New York gossip columns hinting that, as well as Teddy Lynch, Getty might be running around with another of New York's socialites, Gloria Debevoise Spreckels, former second wife of the sugar baron, Adolph Spreckels Jr.

Getty and June had a showdown in April 1938 when she wrote to him angry at his suggestion that a payment of $20,000 should be made to a law firm in Washington to deliver a diplomatic post or to try to induce James Roosevelt to 'use his position to line his own pockets through insurance'. June stated that she would not help her cousin because he had no diplomatic experience and was unwilling to lend her money for developing her dairy business.

Getty's final assault on the Roosevelt administration was made with the help of a gun of a bigger calibre, the California senator William G. McAdoo. First he suggested, in late May 1938, that Getty be appointed to the Civil Aeronautics Board, which was about to be formed. The following month, McAdoo was pressing that Getty should replace Joseph Davies, the US Ambassador to Russia, who was about to be moved to the embassy in Brussels. Getty expressed his appreciation by contributing $5000 to Senator McAdoo's campaign for re-election in 1938 (twice the contribution, incidentally, made by McAdoo's far closer friend, Bernard Baruch). It was money for goods that were never delivered. Roosevelt did not dismiss out of hand a plea from a powerful ally like McAdoo, but neither was he prepared to be rail-roaded into an appointment he might live to regret. Instead, he dropped a note to the Secretary of State,

Cordell Hull, suggesting that they might discuss the
matter. Whether they ever did is not on record, but
Getty's ambitions in the diplomatic field came effectively
to an end until Richard Nixon became President some
thirty years later.

F.D.R. and his advisers undoubtedly made the right
decision, for Getty's political leanings were soon to come
under scrutiny. Between 1937 and 1939 he travelled
around Germany and the rest of Europe extensively.
The Continent was on the eve of war and the persecution
of the Jews in Germany and Austria was already well
under way. If Getty's diary is any guide, the would-be
diplomat wandered through prewar Germany and Austria
blind to Hitler's threat to Western democracies and to
civilization, and oblivious to the torment and ordeals of
the Jews.

He was in Berlin on 26 September 1938, just five days
before Germany marched into the Sudetenland, then part
of Czechoslovakia. He was about to leave for Holland
but instead he went to the Sportspalast in Berlin to hear
Hitler deliver his notorious speech on the Sudetenland to
15,000 of the party faithful.

It is instructive to compare Getty's reactions on that
traumatic evening, as recorded in his diary, with the diary
of another American who listened to Hitler's threat of
war unless Czechoslovakia handed over the Sudetenland.
William L. Shirer, the correspondent of the Columbia
Broadcasting System and author of *The Rise and Fall of
the Third Reich*, wrote: 'Hitler has finally burned his last
bridges. Shouting and shrieking in the worst state of
excitement I've ever seen him in. He stated in the
Sportspalast tonight that he would have his Sudetenland
by October 1st . . . Hitler seemed totally out of control –
fanatical fire in his eye, a nervous tic in his shoulder.'

What Getty saw that night, according to his diary, was

a crowd that greeted the Führer uproariously, and what he heard was 'a speech in favor of peace'. But if he was blind to the events that were unfolding before his eyes, he was still concerned, to the exclusion of everything else, as to how he could benefit from them. He had learned from the Wall Street crash of 1929 that the best bargains were to be had at the moment of the worst disaster.

During a visit to Vienna a few weeks later, on 24 October 1938, Getty, accompanied by a 'typical, charming Viennese' woman, Josephine Bayer, whom he had met in a night club, visited the home of the Austrian Rothschilds at 22 Prinz Eugenstrasse. Since the Nazi invasion of Austria in March, Baron Louis de Rothschild had been held prisoner, while ransom negotiations with Hermann Göring went forward. The Rothschild palace had been looted of its collection of French eighteenth-century furniture, tapestries, silver and paintings. Getty spoke to the police on guard, and 'made inquiry about the furniture', according to his diary.

Apparently, the impulse to obtain that furniture caused Getty suddenly to catch a train to Berlin. Once there, through his great friend, the famous silent-film actress, Charlotte Susa, who had access to high-ranking Nazis like Göring, he could find out what the SS intended to do with the Rothschild furniture. A few days later, over lunch at the Adlon, Berlin's smartest hotel, Charlotte, still an attractive but ageing blonde, 'reported a new law re the Jews; and that the Rothschild furniture may be sold in January [1939]. She will keep me posted,' he wrote in his diary.

Through Charlotte, Getty met a younger and far more sultry actress, Hilde Kruger, at an official party thrown by the Nazi leaders at the Reichschancellery, a grandiose structure built in the style of a Greek temple. Hilde, who

was later the subject of a confidential Federal Bureau of
Investigation file, was born Katherina Matilde Kruger in
Cologne, and was described as a full-bodied blonde with
blue eyes, 5 feet 6 inches tall, weighing 130 pounds. She
was referred to by one junior American diplomat as 'one
of a bevy of gals that were called upon occasionally to
furnish a little "joy thru strength" to the Hitler–Goebbels
combination by night frolicking à la Nero. She was often
in Hitler's company.'

Hilde believed at the time that Getty was 'the guest of
the government' and was negotiating some business 'with
German government officials'. He 'very definitely had
great admiration for Hitler, and surely knew both Hitler
and Göring,' she said in a recent interview. She could
not, however, remember Goebbels ever speaking to her
about Getty.

On 10 November 1938, the so-called Kristallnacht, the
night of the crystal or broken glass, when the smashing of
plate-glass windows of Jewish shops ushered in a week-
long orgy of murder, rape, pillage and arson, Getty was
visiting Lord Duveen's gallery in New York and mulling
over the forthcoming sale of the Rothschild collection.
He decided that most of the furniture was 'too ornate
and massive', but ultimately he bid $72,000 for two
exquisite secretaires which had once been in Rothschild's
palace in Vienna, and was delighted to obtain them at
$128,000 below their value of 1930. So Getty, as usual,
had his bargain.

By the summer of 1939 Getty was back in Europe
during the weeks leading up to the outbreak of war. He
had persuaded Teddy Lynch to train for opera and had
offered to lend her money for lessons. The loan was to be
repaid out of her earnings as a singer. Teddy had
accepted, and Getty chose Blanche Marchesi, the legendary opera teacher, as her tutor in London. Teddy claims

that Getty himself first took several lessons from Marchesi.

That autumn he was in Rome with Teddy and her friend and constant companion, Jean Donnelly. Strangely, Getty stayed at a different hotel from Teddy and there is little discussion of romance in his diary. Rather, he is absorbed in his travels, the purchase of mosaics and ancient busts, his digestion and the price of everything.

Getty had begun keeping a diary again in the prewar years, for the first time since he was a young boy in 1905. He resumed it at the very time he was writing *The History of the Getty Oil Business*, so he probably intended it as a record of a 'great man' who was determined to do even greater things. However, his diaries are mainly a compendium of facts rather than emotions and personal insights. In them he notes the price of travel and records the process of his self-education in art. Women are mentioned as companions at lunch, tea, dinner, theatre, opera and night club. But nothing intimate is ever revealed. If one were to believe the diaries, Getty enjoyed little if any sex life. He often reports reading until the early hours of the morning and appears to leave parties alone, sometimes dining by himself.

Getty did not want the world to know about his sex life, for he told a would-be biographer to leave the sex out as Boswell had done for Johnson. Besides, Getty insisted, he was neither a homosexual nor a drug addict. Late in life he said of the diary entries, 'The thing I notice is the conceit I had. I write of playing with friends and I always push myself to the fore.'

Getty was in Switzerland on 1 September when the armies of the Third Reich stormed into Poland. He heard Hitler's speech in the Reichstag annexing Danzig and noted in his diary: 'Means war. Simply awful. Feel very anxious about T. [Teddy] and J. [Jean].'

Otherwise Getty shows very little emotion in his diary about the war. In the late thirties he was in favour of appeasement, perhaps out of naïveté like many other wishful thinkers. Nor does he comment openly about his well-known admiration for the orderliness and efficiency of Hitler's Germany. On the other hand, his diary entries are hardly anti-Nazi. 'The equipment was superb and the men fine-looking, honest fellows,' he writes about a column of troops he saw returning from the Polish front. He calls 'impressive' a speech by Hitler in Danzig on 19 September in which he hysterically attacked Great Britain.

Getty, an admirer of ancient Roman emperors, expressed no outrage over Hitler's march through Europe. Nor was he ever critical about his friend the Duke of Windsor's pro-Nazi leanings. Rather, Getty remained convinced throughout his life that Britain would have pulled back from the brink of war if the Duke of Windsor had still been on the British throne. By May 1940, however, he was to write in his diary: 'Personally, I am a follower of our F.D.R. and the America First Committee [a right-wing group opposed to the United States entering the war] will get nothing from me.'

Getty's new German actress friend, Hilde Kruger, did not intend to stay in Germany throughout the war. She had been planning for some time to shift the locus of her film career to Hollywood, like Marlene Dietrich. A year earlier Hilde had met the American Ambassador to the Court of St James, Joseph P. Kennedy, and his family in St Moritz, on a skiing holiday. To prepare for her departure, Hilde later went to London to arrange through Kennedy a visa to the USA. On 29 August, two days before war, she left on the last available commercial flight to Oslo. Less than a month later she landed in New York, where immigration officials would not allow her to leave Ellis Island until a wealthy American cotton merchant whom she had met in Germany paid for her landing bond of $500.

Getty remained in Europe. While war raged in Poland, he left Teddy in Rome and returned to Berlin, where he saw a series of women, including Charlotte Susa and Hildegard Kuhn, his old girlfriend from 1930. On Monday, 2 October, Getty took a drive with a woman called Gretchen, dined with Charlotte and made a late date with Hildegard. The next day he added two new German lady friends. He went from one busy social occasion to another. As always, there is no mention of romance or sex. Getty kept careful track of these engagements, but makes no mention of Teddy.

Back in Rome finally by November, he visits the Forum, calling on art dealers, spending his evenings in the luxurious Excelsior Hotel reading, if his diary is to be believed, Gibbon's *Decline and Fall of the Roman Empire*. And there, in Rome, at the Campidoglio, on 17 November, he at last married Teddy Lynch.

Even by Getty's standards this was perhaps the most bizarre of all his weddings. The daily jottings in his diary make no mention of an impending marriage. There is no indication that he and Teddy were having an affair and certainly no mention of love. Before the ceremony, he had negotiated the now familiar prenuptial contract. The only guest at the wedding breakfast was Jean Donnelly, Teddy's best friend.

Getty was planning to return to the USA immediately and hoped to take Teddy with him. He had not reckoned on his new wife's spirit. She told him that she preferred to stay in Rome, where she was now having singing lessons from Jiulio Moreschi, her 'jug-shaped maestro' as she describes him, and would join Getty later. So Getty left alone, with Teddy's words ringing in his ears: 'You dig for oil, I'll dig for my voice.' She commented later, 'He looked like he had been shot.'

Getty sailed the next day from Naples on the *Conte di Savoia* and landed in New York on 23 November. He told reporters who met the ship that he had been in Berlin negotiating a large shipment of oil to the Soviet Union. For the suave international oilman, it must have been disconcerting when government officials suddenly seized his passport.

Mystery surrounds this act by the State Department. Getty was not yet under suspicion for his pro-German leanings. However, he had gained a reputation for trouble with the passport division in the early 1930s by playing fast and loose with the facts in his passport application. Ruth Shipley, who was still in charge of the passport division, knew that Getty had lied on that earlier occasion. Perhaps his most recent passport, obtained on 15 May 1936, was no longer valid, since it had not been renewed for over three years. Several months later, in May 1940, Getty was to tell the State Department: 'I now wish to apply for a new passport and it would facilitate my obtaining it if you would send me my old passport so that I may prove my citizenship.' Getty was probably considering a return trip to Germany because he had left on deposit 47,273.89 registered Reichsmark at the Chase National Bank in Berlin, 7075 Reichsmark at Treuhand-Gesellschaft and 17,915.65 Reichsmark at the Deutsche Bank (worth in total $20,700 in 1939).

10

The Suspect

Heard Hitler's speech and thought it worthy of consideration.

J.Paul Getty

The title of Confidential File No. 100–1202 of the Federal Bureau of Investigation is 'Jean Paul Getty', and it is marked 'Character of Case: Espionage – G'. The file opens with a document dated 26 June 1940, addressed to the US Attorney and signed by a clerk at the Pierre Hotel, New York. He wrote:

To whom it may concern: I am employed at the above hotel. I am an American born here and feel that there should be an investigation made at the above hotel.

About six months ago a few of us heard a conversation which at that time didn't seem very serious, but, as time goes on and these conversations still exist and seem so much more serious at the present time, I think it would be worth while investigating.

There are quite a few Germans and Italians employed here; and very few of our own country employed, so we haven't very much chance to say anything. If you would care for any more information you could get me at the hotel and I would be glad to help in any way.

P. S. Sincerely hope you will take this matter seriously.

It was indeed taken seriously by the US Attorney, Harold M. Kennedy, who sent the letter on to B. E. Sackett, a special agent of the Federal Bureau of Investigation in New York. It was the job of the FBI, in peace or in war, to root out the internal enemies of the USA.

From these modest beginnings began an intensive three-year investigation into Getty's affairs which was fostered by his public support for Hitler's Germany. He invited suspicion with his open admiration for Hitler's leadership, and by publicly bragging of his friendship with leading Nazis, including Hitler, Goebbels, and Göring. He was also reported to be trying to sell oil to Nazi Germany, or to the Soviet Union for shipment to Germany (this was, of course, before the German invasion of the USSR), at the very time when the Roosevelt administration was attempting to win American support for Great Britain. Getty had also invested some of his own money in a German government bond from which he received interest of $60.16 on 17 January 1940, and he was believed to be associating with suspected Nazi agents, including the German actress Hilde Kruger who, by her own admission, had been friendly with Hitler and Goebbels.

At the start of the investigation the FBI was clearly in the dark about Getty. They spelled his name wrongly, calling him 'Mr Geddy', and describing him as 'an enormously wealthy oilman of German descent who comes from California'. The investigation involved agents in New York, Los Angeles and Tulsa, who collected information, interviewed his friends, business associates, former wives and enemies. Information about Getty was also collected by the Navy Department and by Military Intelligence, and was forwarded by J. Edgar Hoover, director of the FBI, to Adolph A. Berle Jr, Assistant Secretary of State, and to officials of the US Justice Department.

Hoover wrote to Berle on 26 August 1940 that 'all the old employees of the Hotel Pierre had been dismissed. This in itself was not so alarming, but it was discovered that the orders for their dismissal had come from Geddy while he was in Germany.' The hotel was said to be 'packed with employees of the Italian consulate'.

Perhaps unaware that he was under investigation, Getty was still trying to obtain an official post in the Roosevelt administration. One day in 1940, after discussing the chance of an appointment at a lunch in Washington, Getty's friend Teddy Hayes returned to the Mayflower Hotel suite which he shared with Edward Flynn, chairman of the Democratic Party. Flynn quashed Hayes's hopes for his friend, saying, 'We got pictures of him and Hitler and Göring; there isn't a chance.' Hayes passed the bad news on to Getty, who responded by writing out 'a check for five thousand dollars to the National Democratic Party'. He evidently hoped that this would balance the unfavourable impression of the photographs.

Getty's problems were magnified by an article in the *New York Daily News* which appeared on 20 December 1940. It described the Pierre Hotel as 'an absolute hotbed' of German and Italian agents and suggested that Getty was involved with a group of Mexicans and Americans in a bid to supply Italy and Germany with Mexican oil.

When the article appeared, Getty was in Mexico attending the inauguration of the new President, Avila Camacho, who was taking over from the revolutionary regime of Lazaro Cardenas, the man who had nationalized all the foreign oil interests in 1938. In Getty's absence, David Hecht, his lawyer, sent a letter to the *Daily News* and to the FBI claiming that Getty had been maliciously libelled. Hecht's letter denied that Getty was a spy, that he knew Adolf Hitler, or that he was trying to sell oil to the Axis powers or Russia. It denied that 'Getty has had business or political dealings' with the Nazis or 'that Mr Getty is pro-Nazi. Mr Getty has been a constant supporter of campaigns and funds designed to aid the British.' On one point Hecht was less than candid. The article had claimed that Getty had returned from Europe in 1939 on the same boat as Hilde Krüger. Hecht rightly denied this, but he went on to claim that they had not met until they were introduced through 'a casual friend' late in 1939. The *Daily News* did not correct its story, nor did it apologize. On the other hand, Getty did not sue the newspaper.

Walter Winchell, the famous columnist, wrote to J. Edgar Hoover on 21 July 1941, defending Getty and claiming that the *New York Daily News* article was inaccurate. Winchell explained to Hoover, according to the FBI report, that Getty could not sue a newspaper, 'since he is a wealthy man and would not have a chance to collect damages'. Nevertheless, other unnamed sources told the FBI that Getty admired the efficiency of the Nazi Government and 'is amiable towards the way those in power crush the weak'. Getty praised Hitler for his 'effectiveness in maintaining discipline in Germany and especially in the army'. It was not until August 1942 that Getty was interviewed by the FBI and claimed 'he had never met' Hitler, Goebbels or Göring. Getty also promised he had 'never publicly expressed himself as being in favor of the things these men stood for'. But, in the same interview, Getty told the FBI that 'he did not have anything directly to do with the Pierre Hotel, because it was owned and operated by the Getty Realty Company, organized by his father.'

Meanwhile, Getty's life in Mexico was becoming increasingly complicated. He was joined there by the explosive starlet, Joan Barry, whom he had met in Los Angeles and who had cabled him in November 1940: 'Need $200. Would you please send it to me.' A few days later she suddenly showed up in Mexico City, professing 'admiration and affection for me', as Getty told the FBI later. He was giving her $150 a month because he was 'impressed at the enthusiasm and energy she put into her work.' She was 'a girl who was on the threshold of having a movie career.'

Getty arranged for A. C. Blumenthal, a notorious *bon vivant* who had been Gloria Vanderbilt's lover, to help Joan gain an entrée to Hollywood film circles. Blumenthal gave her a letter of introduction to a film executive, Tim Durant, who introduced her to his close friend Charles Chaplin. On Joan's first date with Chaplin she made him jealous by telling him that she was still 'in love with J. Paul Getty' and that she had been his mistress for two years and was 'brokenhearted over the affair'. Chaplin became Joan's lover and the relationship developed into one of Hollywood's most tempestuous affairs.

But even if Getty had succeeded in passing Joan Barry on to Chaplin, he was still very much involved with Hilde. The FBI had been on her tail ever since she entered the United States in 1939. They shadowed her on airline flights with her supposed fiancé, Gert von Gontard, a member of the Busch family who controlled a major brewing company in the USA. Every phone call, every telegram sent or received from her hotel room in New York, and most of her mail was monitored. Because she was known as 'Hitler's girlfriend', Hilde could get no work in Hollywood. She made the mistake of being seen publicly, according to the FBI file, in Ciro's Restaurant with Fritz Weidemann, the German consul in San Francisco and an intimate of Hitler from his First World War days.

H. Frank Angell, a young special agent of the FBI, obtained a warrant to search Hilde's bags which were stored at Los Angeles's elegant Beverly Wilshire Hotel. He reported that she had been reading *Mata Hari, Courtesan and Spy* by Major Thomas Coulson and had marked several passages, including one which read: 'Of ordinary spies there are an abundance, but in selection of higher type of spies, those who could move with freedom among men in the highest official circles, their choice was restricted. Mata Hari's well-known reputation as a courtesan showed she had the means of entry to these circles. She was admirably designed for the work.'

Hilde denies categorically that she was a Nazi spy. She maintains, moreover, that some of the accusations made in the FBI file on her are not accurate. She was aware that the FBI was constantly checking up on her, which prevented her from obtaining work in Hollywood, but no one ever proved that she was a German agent.

Nevertheless, early in 1941 Hilde decided that she would try to find work in Mexico, where her Nazi friendships would not be a handicap. According to State Department records, she entered the country stating that 'she intends to marry an American citizen in Mexico in the near future and apply for non-quota immigration visa', but she would not reveal the name of her future husband. The US Ambassador, Josephus Daniels, wired Secretary of State Cordell Hull on 28 February that Hilde was 'reported to be in Mexico in order to join Paul Getty, said to be owner of Pierre Hotel, New York City, and official of Tide Water Oil Company.' The FBI believed that Getty had engineered her Mexican visa, which was true. Hilde became an immediate hit with the Mexican Government. According to official reports, her lovers included Miguel Aleman, the Minister of the Interior, and George Nicolaus, described as 'one of the most active and dangerous Nazi agents in Mexico'. In addition, according to the US naval attaché in Mexico City, 'Miss Kruger is reported to have been a part-time mistress of Raymond Betata (anti-American under secretary of the treasury) and Ezequiel Padilla (rated as pro-American and secretary for foreign relations).'

Getty was evidently not put off. He had given Hilde $5000 to finance her film activities and even asked her to marry him, telling her that he 'didn't care about Teddy' and 'couldn't wait to get a divorce in order to marry again'. He said, 'My only condition is that I want children.' Hilde was evidently put off by his lack of affection and passion. She found him too serious, too prudish, too worried about other people's view of him, and he had 'a very femine attitude'. She believed he was 'scared of women'. Hilde far preferred Hecht, who had 'the most wonderful mind, very amusing, charming and full of life'. Hecht had told Hilde, incidentally, that he was shocked by Getty's admiration for things German. As a Jew, he was embarrassed at having to defend his client's pro-German leanings. Getty makes no mention in his diary of the FBI investigation, or of his controversial relationship with Hilde. Earlier, in Mexico he, Hecht, Hilde and Joan Barry are described as a foursome, but there is no record of what he feels for either woman, except for some annoyance at Joan's harping about money.

However, a confidential national defence informant told the FBI's Los Angeles office that Getty had 'frequently associated socially with Goebbels and Göring during his periods of living in Germany and associated with Hitler' and that he had a common-law wife and two children 'presently residing in Germany'.

Another source, who claimed to have known Getty since he was a young man, was illuminating about his character even if he did not throw much light on his pro-German activities. He told the FBI that Getty 'was a different type of individual than any other businessman he knew; that he was eccentric; that he did not live a normal life as other people do; that he was a mortal physical coward, suffering greatly from an inferiority complex, which he had tried to overcome by seeking publicity, being prominent and getting himself in the limelight as much as possible.' This source, whose name

is blacked out in the pages of the FBI report, added that
Getty 'was the most gullible person he had ever met; that
he would spend hundreds of dollars to attain any sort of
publicity or limelight in fashionable circles amongst
people of noble or high social rank.'

In 1940 Fini told the FBI that Getty had bragged that
he had given the German army advice on how to break
through the Maginot Line. The advice was 'to periscope
[*sic*] under the Maginot Line, inserting a sufficient charge
of dynamite to blow the whole line . . .' Then his fourth
wife, Ann, told the FBI that Getty had been offered 'any
German paintings that he would want' to obtain oil for
them, but that he had turned down the offer.

As for Hilde Kruger, she continued to be the subject
of American surveillance although she did not return to
the United States until the war was over. She was able to
stay in Mexico for the duration without fear of internment
or deportation and protected by high officials of the
government. Articles in *La Prensa*, the Mexican daily
newspaper, also reported her friendship with German
officials and with society figures in Mexico City.

Getty himself seemed able to cross the border between
California and Mexico without benefit of passport. When
he applied for a new one, he claimed that his old passport
had expired on 15 May 1940, had been cancelled by the
Department of State and was mysteriously 'retained at
this Consulate General' (i.e., in Mexico City, where the
application was made). The application, in which he
gave his occupation as 'capitalist', was signed by John
Bankhead, the US vice-consul, and witnessed by David
Hecht. As references Getty gave Thomas Dockweiler,
his lawyer in Los Angeles, and the Honorable J. F. T.
O'Connor, a US district court judge in Los Angeles. A
few weeks later, in March 1941, Ruth Shipley, chief of

the passport bureau in Washington, refused a new passport and referred his application to the fraud section of the State Department. This refusal was repeated on 5 September 1941, on the ground that Getty was 'allegedly pro-German'.

At the same time as he was encountering difficulties over his passport, Getty brought a lawsuit against his mother which might have led a casual observer to conclude that irreconcilable differences had arisen between them. Indeed, Getty had stayed away from California for such long periods that it might be thought that he felt some ambivalence towards his mother. For her part, Sarah was overbearing in her assertions of love. 'My Dearest,' she wrote late in her life, 'you must know that I think of you every hour and wish so much that I could see you.' This time, however, their disagreement centred on the Sarah Getty Trust.

When Tom Dockweiler had drawn up the original trust deed, he had omitted to insert a clause stating that the trust was 'irrevocable'. If it could be revoked, then the assets of the trust, which included a large holding in Pacific Western shares, could be taxed on Sarah's death. When Sarah Getty heard of this, in May 1940, she wrote to her son: 'It is my present desire, purpose and intention, if the trust is revocable . . . to revoke the trust in order that I may be free to use or dispose of the property which will be returned to me.' Getty challenged her decision and won the lawsuit. But the case had, of course, been carefully rehearsed. Both mother and son realized that, if they could persuade the courts to declare once and for all that the trust was 'irrevocable', the future of Sarah's grandchildren and great-grandchildren would be safe. The purpose of the court-room action was to make the trust legally watertight from the taxman.

This is the official explanation of the lawsuit brought

by J. Paul Getty against Sarah C. Getty. However, there is some evidence that Sarah Getty indeed wanted to revoke the trust, to take back her contribution, which had grown in value, and to give the money away as she saw fit. Getty himself, in a deposition taken in 1968, said: 'She had the impression . . . that she had a right to revoke the trust. It was not clear whether she had the right to revoke the trust and she wanted to have it determined.'

In any case, Getty shrewdly brought the action himself on 26 May. The *Los Angeles Times* headlined the family squabble: 'Court Asked to Rule on Whether Mother Can Revoke Notes'. The Los Angeles Supreme Court's speed in setting down the date for five days later, on 31 May, strongly suggests that the judge had been assiduously prepared in the matter. Nevertheless, Sarah had plenty of reason to be disturbed by her son. Getty had not put his son Ronald on a par with his brothers despite the fact that Ronald's maternal grandfather, Otto Helmle, would never make his grandson rich. Helmle's arrest by the Nazis in 1937 for illegal currency transactions had forced him to surrender his fortune to the Reichsbank. He was sentenced to prison for three years and nine months in 1939. Therefore Helmle was in no position to leave Ronald money.

Many years later, in fact, Ronald would argue that Sarah's attempt to revoke the trust was to give him equal financial treatment. Ronald claimed in a lawsuit after his father's death that 'Sarah intended to rescind the Family Trust and establish a testamentary trust under the will that would have equalized [Ronald's] treatment vis à vis his brothers; by promising Sarah that he, J. Paul, would provide in his Will for income to [Ronald] or his issue in an amount equal to that received by his other children and their issue from the Family Trust following J. Paul's death.'

Sarah Getty's lawyer, Rush M. Blodget, Getty's antagonist of several years earlier, was a tough opponent. Blodget told the court that Sarah was still 'independent and rather individualistic'. He argued strongly that the trust was voluntary and could be revoked because there was no evidence that Getty had made any real contribution to it. Sarah Getty never appeared in court to take on her son. She merely testified in writing that the trust was set up in 1934 to 'put the money in notes . . . for Paul to run the business. I meant the property to stay in trust.' The trust, Sarah added in her infirm hand, was 'to protect Paul's children'. Judge Clarence Hanson's judgement, handed down a week later on 6 June 1940, did just that. The trust could never be revoked by any party, Hanson wrote, including 'heirs, executors, administrators, personal representatives'.

It is odd, then, that in 1941 Getty should have slipped through an amendment to a trust that had been ruled irrevocable. This amendment did not make just a minor adjustment. It was major. In the first place a change was put through to benefit Getty over and above his children and grandchildren. The income beneficiaries, of whom Getty was the major one, since he received 80 per cent, would get precedence in any dispute with the so-called remaindermen, his grandchildren. It was never clear why this amendment was put through, but it was done shortly before Sarah died. Even more extreme was the change which for the first time included George Getty II as an income beneficiary to share almost equally with Paul Jr and Gordon. Paul Jr and Gordon would each continue to receive their $9000, and if there was a surplus the three sons – George II, Paul Jr and Gordon – would share equally. It is strange that Getty did not take this obvious opportunity to include Ronald in the trust – and, in a symbolic way, in the family. Ronald continued to be treated as a pariah, financially speaking that is, and would still get only $3000. Getty appears to have convinced himself and his mother that Ronald was still in line to inherit money from his maternal grandfather Otto Helmle, even though he knew that Helmle had lost everything before the war. This exclusion of Ronald was extremely unfair and, after Getty's death, led to a series of lawsuits which even threatened the viability of his estate and the handing on of his fortune.

The court victory over the matter of the trust's irrevoca-bility must have relieved Getty's worries about keeping the fortune intact and untaxed on his mother's death. But he still faced the ongoing FBI investigation into his alleged pro-Nazi leanings. It was a potentially serious embarrassment since it was becoming clear that the USA would not be able to remain neutral for ever.

Just seven months before Pearl Harbor (7 December 1941), the FBI was still reporting that Getty had a close association with Hilde Kruger and that he 'was selling oil to the Nazis' via Mexico. A former employee told the FBI that Getty had agreed to sell 50,000 barrels to Germany by way of Vladivostok, 'but that the deal fell through when Germany was placed on the black list'. Both Japan and Germany were interested in acquiring oil and minerals from Mexico. Hitler sent Dr Joachim Herstlet, a high official, to Mexico to contract for roughly half Mexico's oil production through barter deals. He needed Mexican oil to replace shipments from Russia which had been stopped by his surprise attack on the Soviet Union. From the summer of 1941 the export of all oil products, save those to Great Britain, was subject to strict controls.

Hitler desperately needed imports of oil in order to

fight a protracted war in Europe. He had been able to swop industrial products for Soviet oil and there were rumours that American oil companies had been shipping crude oil to the Soviet Union which could be trans-shipped to Germany. Getty at one point had contended that 'Russia had all the oil she wanted in Europe and if she wanted to sell any to Germany, that was her business.' Churchill's Ministry of Economic Warfare did not agree. It complained to the Roosevelt administration that 'a good deal of crude oil is sent to Vladivostok from the United States, releasing certain amounts for export to Germany.' Until the summer of 1941 the USA also supplied Japan with substantial amounts of oil to prevent her from invading nations with oil in Southeast Asia.

Getty may not have been selling oil from the USA or Mexico to Germany by way of Russia, but he was using his high-level Mexican connections to acquire a choice piece of beach property on the Pacific Ocean just south of Acapulco. A large portion of the land bordering the Pacific consisted of ranches owned by an American family called Stevens who had made the mistake of not placing their property in a Mexican company. Therefore it was vulnerable to expropriation and Getty saw an opportunity to take advantage of their predicament. With the help of leading Mexican politicians like Aleman, Getty proceeded to negotiate the cheapest property deal he ever made. Hal Seymour, Getty's cousin, who was left in Mexico to ride herd on the transaction, wrote to him during the summer of 1941: 'Nobody BUT YOU will own one metre of that land.' The average cost per acre was 6 cents, according to Stuart Evey, a Getty oil executive who looked after the property in the sixties.

July 1941 found Getty in Santa Monica, writing the introduction to his opus *Europe in the Eighteenth Century* (privately published in 1947). 'We of the twentieth century', he wrote, 'have been profoundly influenced by the manners, customs, philosophies, politics and arts of the eighteenth century.' Portrait painters like Gainsborough, Reynolds and Romney were 'part of the aristocratic life of the days before democracy, anarchism, socialism, communism, fascism and nazism. Their subjects have the calmness and sureness, the leisurely grace and courtly manners of a bygone age.'

Suddenly and surprisingly, in September 1941 Getty resigned as sole trustee of the Sarah Getty Trust and named Thomas Dockweiler to replace him. He explained later that he did so in order to devote himself to the war effort, but his resignation again reveals his curious ambivalence towards his responsibility to the family business. However, in a diary entry for 15 September 1941 he wrote: 'A trusteeship is a job for an attorney or a bank or a group, not for one layman.' Seven years later, when George II became a trustee and opposed him, Getty was to have a definite change of heart.

Just prior to his resignation Getty attempted to merge Pacific Western and George F. Getty Inc. on terms favourable to the Getty family holdings, but a small stockholder brought a lawsuit complaining about the terms and stopped the transaction.

At the same time, as the war clouds blew more fiercely, Getty was trying to obtain a commission in the US Navy, although, he wrote later: 'I had at first hoped that the United States would be able to stay out of the war. And like so many of my fellow countrymen, I grew to realize that no free nation could long continue to exist if the remainder of the world was controlled by totalitarian powers.'

After Pearl Harbor, J. Edgar Hoover personally issued

approval for the custodial detention of Getty as a potential enemy. However, he wired his Los Angeles office a
few days later that Getty, and several other individuals,
were not to be brought in yet. This was fortunate for
Getty because, late in 1941, his mother Sarah, then in
her eighty-ninth year, contracted pneumonia at her home
on South Kingsley Drive, Los Angeles, and on the day
after Christmas, seven years after the trust in her name
was formed, she died. Getty was genuinely heartbroken
and for days afterwards his diary contains continual and,
seemingly, affectionate reference to his mother: 'How I
miss her . . . Mama, Mama . . . Mama is gone.'

A devoted member of the Women's Christian Temper-
ance Union, Sarah Getty had lived in the same Tudor-
style mansion for thirty-six years. Her obituary, perhaps
suggested by her son, claimed that her husband George
had been 'at one time the largest independent oil producer
in the nation', a claim that might well have been contested
by several wily Texans such as H. L. Hunt and Clint
Murchison.

Sarah made it clear in her last will and testament that
there must be no opposition to her bequests. Anyone
challenging her wishes would be treated 'as if such person
never existed'. George Getty II was left the house on
South Kingsley, and J. Paul was forbidden to use the
mansion as security for a loan. Ronald was left $200,000
by way of compensation for his unequal treatment by the
trust. Paul Jr and Gordon each received $50,000. If these
four sons of Getty had no children, Sarah instructed
that the trust should be used for the 'advancement and
promotion of the fine arts'. A perpetual fund, known as
the Getty Art Foundation, would be formed to pass
income on to the Los Angeles Museum of History,
Science and Art.

The close of 1941 must have been one of the most

emotional periods of Getty's life. 'Now I am an orphan,' he wrote in his diary. When his aide Staples came to pay his last respects, Getty, with tears in his eyes, said, 'She was my last connection.' Moreover, his wife Teddy, whom he had not seen since his wedding day more than two years earlier, had been arrested by Mussolini's police on 11 December, the day that Italy declared war on the USA. She was 'suspected of political activity', according to the State Department. Secretary of State Hull sent a telegram to Getty on 22 January 1942 stating that Teddy 'was unable to obtain plane transportation to Lisbon before the outbreak of hostilities'.

So, ironically, while Getty was under investigation in the USA for being a suspected Nazi sympathizer, Teddy, still in Italy studying music, was arrested for being an American spy. She was released on parole, then re-arrested and thrown into prison for several days. There is no precise explanation for Teddy's second arrest. Because she had credentials with the *New York Herald Tribune* she was again released and sent with a group of American newspapermen to Siena where they remained in a small hotel until the State Department made arrangements for their passage aboard a neutral Swedish liner in mid-1942. During this period Teddy was able to send messages back to Getty by way of the State Department.

Getty's efforts to get a commission in the United States Navy failed. In his autobiography he claims that this was because his sight was bad, and adds that Navy Secretary Frank Knox, a former Chicago newspaper publisher and a friend of Bill Skelly, told him he could best serve his country by running the Spartan Aircraft Company at Tulsa. This was a subsidiary of Skelly Oil which made parts for aeroplanes and also ran a training school for fighter pilots. There were several versions of Getty's failure to join the Navy. In addition to his own reason of defective eyesight, Skelly's son-in-law, Harold Stuart, believes he was fearful of being drafted, and that was why he 'wanted Skelly to get him a commission in the Navy'. Army Intelligence, reporting in 1942, had another version: 'Subject [Getty] supposedly used Mr Skelly's influence with Colonel Knox to obtain for subject a commission in the Navy, but later lost it; according to rumor, due to the fact that subject insisted upon some guarantee of shore duty and, being unable to obtain such, resigned.'

And so, by late February 1942, Getty had returned to Tulsa, a vastly different place from the rough boom town he had left in 1916. Even as he was settling down at the Spartan Aircraft Company, a confidential memorandum prepared by Naval Intelligence was sent to the State Department. It repeated all the old stories about the Pierre Hotel, the sale of oil to Germany, the boasts about knowing the Nazi leaders and Hilde Kruger. Dated 22 March 1942, the memorandum did include one new revelation. Getty's application for a commission as a naval officer was not turned down because of age or eyesight. It 'was rejected because of suspected espionage activities'. The 8th and 11th Naval Districts considered '*SUBJECT* to be dangerous.'

11
The Patriot

All I knew about airplanes was that they had wings and engines
– and that if they were properly built and piloted they flew.
                                                    J. Paul Getty.

The new president of the Spartan Aircraft Company
occupied a four-room blockhouse which he had had built
for him. Located directly beneath the main flight path of
Tulsa municipal airport, it had 12-inch-thick walls above
ground and 18-inch walls below ground, and housed a
marble fireplace on which was carved the nose-end view
of an aeroplane. The building was, in fact, a concrete and
steel bunker, designed to avert the three dangers which
Getty feared most – the direct crash of a loaded bomber,
the vicious tornadoes of the Midwest and, even more
exaggerated, the effects of a bombing attack by the
German Luftwaffe. Although he had also bought a house
in the residential area of Tulsa, this fortress was to be his
office and his home and he seldom ventured outside its
safety. Here he settled down, as he recorded in his diary,
'to do my best to be worthy of Mama and to help
my country crush its enemies to the last ounce of my
strength.'

The Spartan Aircraft Company of Tulsa, Oklahoma,
had been formed in 1928 by Bill Skelly, who was fascinated by the future of air travel. It built one or two civil
aircraft, but the troubles faced by Skelly Oil during the
1930s meant that Spartan's business got little attention
and by 1941 it was not too healthy. It made rudders for
the B-24 Liberator bomber and engine mounts for the

P-47 fighter, and it also had a lucrative contract with the United States Navy to build pilot training planes, codenamed the NP-1. Unfortunately this aircraft was not considered safe for solo students – it was alleged to have substandard welding and could disintegrate in a tail spin; two weeks before Getty moved to Tulsa, the Navy had grounded ten of the machines and threatened to cancel the contract. Now they decided to give the company a last chance.

More successfully, Spartan also owned a flying school for pilots from the United States Air Force, the US Navy and the British Royal Air Force, run by Captain Maxwell Balfour, a veteran flyer who had seen action in the First World War. The largest aeronautical school in the world, it would turn out 15,000 American and Allied pilots between June 1939 and early 1944 and make $2 million profit.

Before the war, Spartan Aircraft had been run by its founder, Bill Skelly, who was also president of Skelly Oil (its parent company) and of Mission Corporation, which owned 57 per cent of Skelly Oil. Thus, when Getty gained control of Mission, he had also taken over the aircraft company, to which he had not paid too much attention. But now he took over the day-to-day management, ousting Skelly in the process, to the latter's mortification. He showed his resentment by keeping an eye on Getty's activities throughout the war and ensuring that Spartan Aircraft's book-keeping and government contracts were checked by Skelly personnel.

At no other period in his life did Getty work harder or more unselfishly than now, when he made a religion of the war effort. He transformed the stock market operator and art collector into a novice factory manager, responsible for shop design, tooling, engineering and personnel matters. 'The secret of success for a sub-contractor like

Spartan is to meet the prime contractor's schedules promptly,' he wrote in his diary. 'Excellence of design and salesmanship are not factors.'

He stayed at the factory until all hours, skipped lunch, and paid himself a salary of only $100 – in nickels, which he used in the Coca-Cola machine. He drove the workers hard too. On 30 March 1942, shortly after he took over Spartan, he asked the factory to work without pay on a Sunday to make a free trainer for the Navy. It was an excellent public relations gesture in view of the threatened loss of the naval contract. At the presentation, Getty told the workmen: 'You are loyal, honest, hardworking American citizens – people to be proud of – people to pattern by.' He emphasized that Spartan 'was an honoured name in ancient Greece over 2000 years ago. Let's keep it so and make it even more glorious.'

As the president of a comparatively small manufacturing company, he cut quite an eccentric figure, living as a virtual recluse in his concrete bunker and not socializing. Walter Bishop, the company's purchasing agent, recognized his passion for 'not being wasteful. He wanted to make use of every piece of material and didn't throw money around.' His mania was efficiency and more efficiency. 'He demanded it and he got it,' says Bishop. When he was away from Tulsa each manager had to write a daily log of all his activities and these were given to Getty at the end of the week, thus fulfilling his passion for day-to-day control.

'I like to think I have made a worthwhile contribution to America's war effort and without any thought or possibility of financial profit,' he would confide to his diary in 1944. But when he originally took control of Spartan Aircraft not everyone was convinced about his newfound patriotism. Ever suspicious, J. Edgar Hoover asked the Office of Naval Intelligence to investigate Getty's activities at Spartan.

Although some informants claimed that the company employed people with pro-Nazi sympathies, the Navy's report on Getty himself concluded: 'It would appear, lacking other evidence, that Getty has been indiscreet in his choice of associates and naïve in his interpretation of the political scene, rather than an avowed supporter of the Nazi or Fascist regime.' It had taken three years of investigation to show that he was not an enemy spy, but merely an admirer of German, and especially Nazi, efficiency. His many enemies, coupled with the incompetence of the FBI, had created a record that was to dog him for the rest of his days. Every single President after Roosevelt ordered a top aide to read Getty's FBI report.

At the same time the War Department was conducting its own investigation. A military intelligence officer reported in June 1942, that Getty 'works at odd hours and into early hours of the morning; has irregular habits; leaves on unannounced trips for several weeks at a time; has few friends in Tulsa, if any, keeping company mostly with [those] whom he employs or imports as house guests, very often persons of doubtful character.'

On one point the two investigations agreed. The War Department said: 'Subject reputedly disorganizes and upsets every organization to which he had gained access.' That sentiment is echoed in the Navy's report, where he is described as 'a financial genius at obtaining control but thereafter a genius at disorganization.'

As we have seen, he was equally a genius at disorganizing his private life. His wife Teddy at last arrived in New York in June 1942, aboard a neutral passenger ship, the *Grispholm*. Interviewed on disembarkation, she told the waiting newspapermen: 'Mussolini did a great deal of good, cleaning up the place, excavating things and so on.

Then afterwards, the spaghetti got darker and Ferragamo
stopped making those wonderful shoes.'

Having thus summed up the pros and cons of the
Fascist regime, Getty's fifth wife proceeded to Tulsa
where they spent their wedding night two and a half
years after the wedding. But the bridegroom, who had
not set eyes on his bride since their wedding day, did
move into his house in the residential quarter of Tulsa for
the occasion. After a short period in residence, however,
Teddy moved into the beach house in Santa Monica,
formerly occupied by Ann, to resume her singing career.

This left Getty free, in so far as his work left him with
any free time, to carry on affairs with his dentist's
secretary and with waitresses he picked up in the Hotel
Tulsa, one of his old stamping grounds. It had to be done
surreptitiously, for Tulsa was a small conservative city
where social conformity was expected. So Getty might
not have been pleased when the voluptuous Joan Barry
showed up again in Tulsa in late 1942. Having taken her
under his wing, Charles Chaplin had bought her a silver-
fox coat for $1100, enrolled her in the Reinhardt school
of acting and put her under contract at $250 a week to act
in a film of the play *Shadow and Substance* of which he
had bought the film rights. But this 'big, handsome
woman of twenty-two, well-built, with upper regional
domes immensely expansive,' as Chaplin described her,
was also highly strung and emotionally unstable. Often
drunk, Joan Barry would yell abuse at Chaplin and their
stormy relationship provided a rich diet for the gossip
columnists.

She suddenly announced that she no longer wished to
be an actress and said she would tear up her contract if
Chaplin gave her $5000 and paid her fare and that of her
mother to go back to New York. Chaplin was happy to
agree, but instead of going to New York she arrived in

Tulsa, threatening suicide and asking Getty for money. Through his Tulsa attorney, Claude Rosenstein, Getty agreed to lend her $1674.80. Security for the loan was afforded by her automobile and the silver-fox coat.

She did not stay long in Tulsa. In December 1942 she returned to Los Angeles and entered Chaplin's house armed with a gun. Somewhat surprisingly, in the light of this dramatic entry, she spent the night in Chaplin's bed. Her subsequent behaviour, however, when she kept turning up at Chaplin's house uninvited, often drunk, finally extinguished his interest in her. In January 1943 he had Joan arrested for vagrancy. She was sentenced to jail for ninety days, but was told she would not have to serve the sentence if she left Beverly Hills and did not return for two years. Once again Chaplin agreed to pay her fare to go back east and once again she turned up in Tulsa in order to get more money from Getty. He hired a private detective, F. N. Burns, to keep an eye on her. She passed a bad cheque in Deidenbach's department store and refused to pay her bill at the Mayo Hotel. Getty told the private detective to pay both the bills to avoid a scandal. She returned to New York, staying in Getty's hotel, the Pierre, where she attempted to commit suicide. On 16 April 1943 Getty made a last loan of $700 to Joan, insisting that she sign an agreement promising never to ask for money again and 'to the best of my ability to contribute of my talents and time to the successful prosecution of the war in which this nation is now engaged.' This was a somewhat amateur attempt, it might be thought, to emulate the British Prime Minister William Ewart Gladstone in his efforts to reform London's ladies of ill repute.

Charlie Chaplin was now to come under attack from two sources, Joan Barry herself and the FBI. In June 1943, shortly before he married Oona O'Neill, the charming seventeen-year-old daughter of Eugene O'Neill, Joan Barry arrived once again unannounced at his home and declared she was three months pregnant. The police were called and she was evicted, but to the press, who had been alerted, she gave a suitably hysterical performance as the abandoned waif with whom Chaplin had had his wicked way. A week later she served him with a paternity suit.

Meanwhile the FBI saw their opportunity to get at Chaplin. They had first been alerted by right-wing isolationists and Nazi sympathizers when he made his hilarious anti-Hitler film *The Great Dictator*. The FBI were as thorough in investigating Communist sympathizers as they were in rooting out pro-Nazis, and their suspicions had been confirmed when, along with other actors, including Orson Welles, Chaplin attended a public meeting in Carnegie Hall, New York, to raise money for Russian war relief and spoke out in favour of a second front.

Joan Barry, who was staying at the Pierre Hotel, attended the meeting accompanied by Getty's trusted lawyer, Hecht, who knew her from their days in Mexico. After the meeting, he escorted her to the Stork Club. The FBI now sought to use Barry as the main line of their attack on Chaplin. They interviewed Hecht, who told them Joan was 'nervous, temperamental, shrewd and calculating'. They interviewed Getty, who insisted that he had never had any sexual relations with her. (But he was to write later, in his autobiography, 'Miss Barry, an aspiring actress, had been one of my girl-friends for a brief (and I must confess) quite hectic period.') He told the FBI agent, 'As far as I know the girl has always been well behaved.'

Joan herself told an FBI man that Getty 'has an interest in actresses, but that interest does not include being intimate.' She told a Tulsa policeman, however, after she had been arrested for passing another bad cheque, that she did indeed sleep with Getty.

Armed with this evidence, or lack of it, the Justice Department now began a somewhat ludicrous investigation by a grand jury in an attempt to prove that Chaplin had violated the Mann Act when he took Joan Barry to New York at the time of the Carnegie Hall meeting. This was the Act prohibiting the transportation of women across state lines to sell their bodies for immoral purposes.

So Chaplin had to face two public trials – first a civil suit for making Joan pregnant, and the second on two counts under the Mann Act. He was also indicted under a long-forgotten law on two further counts, which alleged that he deprived Joan Barry of her civil rights when she was told not to return to California for two years.

Until the paternity case came to trial, Chaplin was forced to support her child. Even though a blood test proved that he was not the father, he became her chief financial support, giving her $10,000 for the child's maintenance during the first year of its life. Chaplin was acquitted, although several jurors had thought the child was his, according to an interview with a jury member, Ferdinand J. Gay, in the *Los Angeles Times*. But he added that the other seven jurors 'could not overlook the fact there was testimony showing she had been alone with Getty and Ruesch [another of her lovers] around the time she says she was with Chaplin.' There is no doubt that she was in Tulsa in the middle of January 1943 and the baby was born on 2 October 1943, some nine months later. Getty denied sleeping with her.

And now Getty had to appear publicly and testify at the federal government's trial against Chaplin on the ridiculous Mann Act charges. Luckily, the judge in the case happened to be his very good friend, J. F. T. O'Connor, whom he had given as a reference when he applied for a passport in Mexico in 1941. It was a still greater coincidence that O'Connor had met Joan through Getty and, after spending an afternoon with her, the Comptroller of the Currency had given her a copy of his book *Banks Under Roosevelt*, inscribed: 'To Joan Barry, with kindest regards from her friend, the author, J. F. T. O'Connor, Los Angeles, September 2, 1942.' It was perhaps surprising that the judge did not withdraw from the case, but Joan denied being sexually intimate with him when the question was raised.

Getty was treated with kid gloves by his friend when he had to take the witness stand. The judge instructed him to answer simply yes or no to questions about his relationship with Joan. He never had to give any details revealing his own deep involvement which might have aided Chaplin's defence.

The government could not prove that Chaplin sold Joan's body to others across state lines any more than Joan could prove that he was the father of her child. There is no evidence that Getty was the father. We do know that Joan received $3688.53 in all from him, which was less than the nice round sum he had given to Hilde Kruger. He explained to the FBI that he had practised this personal form of philanthropy with about a hundred individuals over a period of twenty-five years. 'If I had all the money I have lent to people in the last twenty-five years,' he said, 'it would amount to a very large sum.'

In a series of articles syndicated to the national press under the title 'The Mystery Man of Tulsa', he denied that Joan 'had ever tried to blackmail me', but he must have been deeply concerned at the danger she posed to his image as a committed patriot fighting the Second World War from his bunker. On her last visit he arranged through his lawyer to have her escorted out of Tulsa by two detectives on an aeroplane.

* * *

As the war began to draw to a close, Getty turned his attention to the task of converting the Spartan Aircraft Company to the manufacture of peacetime products, saving several thousand jobs in the process. 'Damned if I'll just blow the whistle and have the Spartan factory turned into an ice-skating rink,' he wrote in his diary in mid-1945.

Together with Captain Balfour, he examined the possibility of manufacturing many different products including home appliances, refrigerators, even automobiles. At one point he proposed that the Spartan Executive, a small propeller-driven aircraft, might become the nucleus for a civil aircraft company like Beech, and dreamed of supplying passenger planes for 'a great network of feeder airlines to cover the southwest'. He even prepared a brochure that promised 'the name Spartan will be synonymous with the highest quality aircraft products'. But he decided not to go ahead and risk a great deal of capital competing with established companies like Cessna and Northrup. However, he did negotiate the purchase of Spartan from Skelly in a classic Getty deal, involving a down payment of only $50,000. The rest of the purchase price of $1,441,546 was in the form of a promissory note.

Meanwhile the veteran pilot, Captain Balfour, came up with the idea of manufacturing mobile homes, which he had seen under construction in Detroit. By early 1946 Getty was totally absorbed in Spartan motor homes, personally checking every complaint from the 2169 trailer buyers in 1947. He showed the same fanatical interest in the minute details of his business that he had always shown in his personal life. Marjorie Garrison, his secretary, had to check every item in his grocery order before he would pay for it, and throughout his life Getty personally washed his underwear every evening, because he did not trust the detergent used in commercial washing machines.

In spite of his successful forecast of America's demand for cheap housing, even housing that you could take with you from coast to coast, Spartan was up against stiff competition. It lost $2 million the first year, bringing Getty's total investment to $3.4 million, and although it showed a profit of $500,000 in 1947, Getty was losing interest. He became distracted again by many other possibilities in his life, including another retirement.

He became a father for the fifth time on 15 June 1946, when Teddy gave birth prematurely to Timmy, who weighed only 4 pounds 14 ounces. Getty told his secretary to send roses to Teddy and 'something for the baby, maybe a little hat'. It was three weeks after Timmy's birth before Getty went to Los Angeles to see Teddy and his child at the hospital.

Teddy was good at giving advice on how to ensure a successful marriage. 'Don't hold him on a tight rein,' was her counsel. 'Treat him like you would a good horse. Hold the rein loosely, but get a good grip, so he knows he's being held.' In fact her grip was extremely loose. Marian Anderson, the statuesque former swimmer and beauty queen, used to appear at the beach house in Santa Monica early in the morning, according to Ralph Hewins's biography of Getty. Belene was still a fixture too. Amazingly, Teddy stuck to her singing career, even landing a role in the film of *Lost Weekend*, and Getty set her up in the unlikely business of bottling natural spring water from Texas. What had begun as a bizarre marriage continued on its eccentric course.

Even when he was in Santa Monica, Getty spent little time with his family, according to Maxie Sickinger, a German physical fitness enthusiast who looked after the paraphernalia at the Santa Monica Beach Club. Maxie would walk with an unshaven Getty on the beach and listen to him as he declared his ambition to become a beachcomber. He never attended the parties at the Beach Club with Teddy, and did not even seem to be interested in Howard Hughes, an even greater eccentric, who lived for a while in the penthouse at the club.

At fifty-five, Getty was no longer the lean, mean playboy of a decade earlier. With slightly stooping shoulders and greying hair, now dyed reddish brown (he had used blonde dye in his early thirties), he underwent his second facelift (the first was in 1939). But the operation seems not to have made him more attractive and younger looking, but rather older with pinched features. The effect of the facelift, and the ones to come, was to accentuate the apparent size of his already large nose and his elephantine ears.

During the war Getty had left the direction of the Getty oil holdings and the Sarah Getty Trust to others. The management of Pacific Western and the Getty oil leases was handled by Tide Water, the company he had tried in vain to capture. In 1946, to simplify the corporate structure, Getty finally merged the private Getty oil holdings into Pacific Western. This meant that his holdings could be more easily evaluated because Pacific Western was a publicly traded company. But the postwar stock market valued Pacific Western shares at only $21.75 each, far below their value in terms of the oil reserves Getty controlled. At $21.75 each, Getty's Pacific Western shares were valued at a total of $9,787,500, while the Sarah Getty Trust was worth slightly over $15 million. However, through consolidation, Getty now owned 34 per cent of Pacific Western, while the trust owned 51 per cent.

Apart from a soggy stock market, Getty could find several reasons to sell his oil holdings in order to fulfil his fantasy of becoming a beachcomber. The federal government had singled out Pacific Western in its attempt to obtain tax revenues from the oil found in the California tidelands just off shore. If the Truman administration were successful, it would mean far more federal control over a major source of Getty's wealth. The private publication of *Europe in the Eighteenth Century* signalled Getty's need to be a writer, and also perhaps his need to return to Europe, where the monuments of the eighteenth century awaited him.

Nevertheless, it is strange that Getty should have considered selling out just as the price of crude oil had climbed from $1.19 a barrel to $1.93 in 1947 and seemed to be moving over the $2 mark for the first time since 1919. Moreover, it would be an admission of failure in the quest to take control of Tide Water.

As with his rejection by the Iraqis in 1931, Getty never voluntarily revealed the most important turning point in his entire business career – his defeat by oilman Bill Skelly in an attempt to sell the Getty oil holdings for cash. There is no mention of this episode in either of the two autobiographies, his official biography or dozens of articles and interviews.

Getty must have known he needed to pull a ruse on Skelly, who was still president of Mission Corporation, which owned 55 per cent of Skelly shares. Secretly, Getty's chief aide, David Hecht, negotiated to sell Getty's and the trust's Pacific Western shares at $68 a share to Sunray Corporation, another Tulsa-based oil company. This would have meant a substantial profit for Getty, whose average cost for the Pacific Western shares was slightly over $4 a share. The transaction would have given Getty and the trust almost $80 million in cash, some seventeen times the asset value in late 1934. Then there was to have been a three-way merger of Pacific Western, Mission and Skelly into Sunray for shares of Sunray stock. Getty knew that Skelly would take a punch at anyone who crossed him. A stocky, fierce-looking man, with a balding pate and furrowed brow, Skelly resembled the bust of a Roman general Getty had once seen in a Naples museum. But Getty did not realize how ferociously Skelly would struggle to stop Clarence Wright, the president of Sunray, from replacing him as 'Mr Tulsa' or the oilman most prominent in civic affairs.

At a Mission board meeting on 18 October 1947 Skelly voted against the merger with Sunray and was thrown off the Mission board by a vote of Getty's aides. The strongarm tactic did not deter Skelly. He went to court in Reno, Nevada, on 4 November 1947, determined to stop Getty. Skelly argued correctly that the terms of the deal did not value Skelly's oil reserves fairly. Moreover, Getty had not taken time to obtain an independent valuation of Skelly's assets to prove that the merger terms were fair. It came as a shock when it was disclosed that Sunray had also planned to sell Mission's block of Tide Water to raise money to buy out Getty's Pacific Western stock.

On 3 December 1947 Judge Roger Foley held up the merger because its terms were 'unequal' and favoured Getty interests over Mission shareholders. Getty appealed, but on 12 December the court in San Francisco upheld Judge Foley's injunction. Ten days later Getty abandoned the transaction. Magnanimous in defeat, he told Skelly: 'Every time I've crossed you, you have beaten me. You run the oil company.' And in fact, as president of Skelly Oil, Bill Skelly doubled its net profits in 1947 and continued to hit new records for crude oil production every year.

What Getty did not tell Bill Skelly was that he intended to go ahead anyway and sell the Getty holdings in Pacific Western to Sunray at $68 a share, even if the merger was off. The Gettys would still get their $80 million.

This time the opposition came not from Bill Skelly but from Getty's own eldest son, George, who had joined Dockweiler as trustee of the Sarah Getty Trust in July 1946, when he was twenty-two. Young George, who had been brought up by his mother, Jeanette, and his affectionate stepfather, Bill Jones, a Los Angeles stockbroker, had hardly ever seen his real father, but he felt a compulsion to join him in the oil business. His power over the trust gave him the right to do the unthinkable – oppose his own father's wish to retire. He must have shocked his father when, on 11 January 1948, he went to court to try to change some of the terms of the Sarah Getty Trust. He claimed credit for pushing Sunray's bid to $68 a share, and thought the price should be even higher. He also wanted to change the trust's restriction about investing only in the bonds of the USA, Great Britain, Sweden, Denmark, Norway and Switzerland, and to be able to put the money in real estate.

Getty put down the rebellion in a forthright manner. He brought his own legal action asking to be reinstated as trustee in the place of Dockweiler and George, and admitting that his 1941 resignation had been a mistake. He claimed that there was disagreement between Dockweiler and George Getty, and persuaded the court that the shares of Pacific Western should be in the hands of an experienced oilman rather than two trustees 'where there is a difference of opinion'.

Getty said in a court deposition in 1968: 'Tom Dockweiler was another case more like my mother. I mean he was ultraconservative and the companies were making no progress. George came in as a trustee but he was young and inexperienced . . . and I think it was felt that it was better for him to try his hand as an individual oilman rather than go into a large oil company.'

George acquiesced and resigned, and by 21 May 1948

J. Paul Getty was back in sole control of the Sarah Getty Trust. The very next day, he registered his personal holding of 450,227 shares of Pacific Western with the Securities and Exchange Commission, a legal requirement for anyone wishing to sell part or all of a controlling block of shares. He indicated that he would offer the shares for sale from time to time on the floor of the New York Stock Exchange.

Unfortunately, the public squabble between father and son, between the head of the Getty oil empire and one of the trustees of the Sarah Getty Trust, was beginning to backfire. Sunray had no wish to get involved in the quarrel. When the lawyers for Sunray's bankers intimated that Getty might very well not have the legal right to sell the trust's stock at all, but only his own, leaving them with only a minority interest in Pacific Western, they decided that it was not worth the hassle and backed down. Concrete proof that Getty wanted to sell out was the disposal of 15,400 shares of Pacific Western stock in 1948 for a profit of $722,127.54. It is the only known time that Getty sold part of his holdings.

And so, in the end, the language of the trust, dictated by Getty's mother, prevailed. It proved to be the final obstacle and by July 1948 the Sunray deal was totally dead. Getty had been forced to stay in the oil business. If he wanted 'to travel and see the world again', as he said in his diary, he would have to do it as the executive of an oil company, not as a carefree playboy. But within a decade the $80 million's worth of Pacific Western stock which he had wanted to sell was to multiply again and again until, in 1957, the magazine *Fortune* declared that J. Paul Getty was the richest man in America.

12
The Gambler

The meek shall inherit the earth, but not the mineral rights.
J. Paul Getty

Frustrated in his attempt to sell his oil interests and retire, Getty now decided to expand and to create the worldwide oil empire that had once been his goal. The postwar American economy was beginning to boom and ribbons of federal highways were being built with government funds to cope with the growing flood of automobiles. America simply could not meet its own insatiable oil needs and must grow increasingly dependent on foreign oil. That meant Mid-Eastern crude, and the first shipment of Arabian light crude had been landed on the east coast in May 1948.

Many oilmen believed in the 1950s that all the big oilfields in the United States had been discovered, and Getty's own Pacific Western Oil Company had made no major new finds for over a decade. It owned 500 small wells, producing a mere 9000 barrels a day, an average of only eighteen barrels per well. The obvious place for him to find a huge increase for his companies' reserves was the Middle East, but that meant that he would have to break down the barriers erected by the seven giant companies which dominated world production and transportation of oil and were known collectively as the Seven Sisters. These were the Anglo-Iranian Oil Company (later to become British Petroleum), Royal Dutch/Shell, Standard Oil of New Jersey, Standard Oil of California, Socony Vacuum (the forerunner of Mobil), Texaco and Gulf.

The Seven Sisters combined in groups to exploit the vast oil reserves of the Middle East. They controlled more than 80 per cent of all Persian Gulf production and dominated the political and economic structure of the area. In Saudi Arabia, four American giants – Standard Oil of New Jersey, Texaco, Standard Oil of California and Mobil – combined to form the Arabian-American Oil Company (ARAMCO) and pumped out 525,000 barrels of crude oil a day. In Iraq, Standard Oil of New Jersey and Mobil were partners in the Iraq Petroleum Company. Iran was dominated by British Petroleum, half-owned by the British Government. Independent oil companies barely mattered in the Middle East at the time, but Getty knew that 'if one is to be anybody in the world oil business, one must have a footing in the Middle East.' Like every other independent oilman, he had read the Texas geologist Everett DeGolyer's report to the Truman administration which concluded 'reserves of great magnitude remain to be discovered' in the Persian Gulf area, 'the center of oil production'.

Getty was determined not to lose this chance of gaining a foothold in the Middle East. Now he had his sights firmly fixed on one smallish area which the oil giants had ignored, the so-called Neutral Zone. This was a wedge-shaped piece of desert, 2200 square miles in area, bounded on the north by the kingdom of Kuwait, on the east by the Persian Gulf and on the south by the kingdom of Saudi Arabia. It was an expanse of desolation, the windswept sands broken only by the occasional outcrop of rock, the landscape sustaining neither beast nor vegetation, its only visitors the nomadic Bedouin tribes who know no frontiers. Yet, for all the apparent worthlessness of this piece of Arabian real estate, the Saudis and the Kuwaitis had squabbled for centuries over who owned it; the issue was settled with a typically Arab compromise.

Nobody did. It would be deemed a neutral zone, a kind
of peacetime no-man's-land. However, Saudi Arabia and
Kuwait would exercise dual sovereignty over its mineral
rights. If the question of a concession in the area cropped
up, Kuwait and Saudi Arabia would each grant it. If each
government chose a different oil company or consortium
to drill in the area, then whoever found oil would have to
share it with the other concessionaire. In effect, the two
concessionaires would be forced to become partners,
whether they liked it or not.

Just a few miles north of the Neutral Zone's vague
boundary lines lies the greatest oilfield in the world, the
giant Burghan field in Kuwait. There, British Petroleum
and the Pittsburgh Mellons' Gulf Oil had combined to
form the Kuwait Oil Company. Already they had found
11 billion barrels of high-quality light Arabian crude,
filling the coffers of the royal family of Kuwait with cash
in the process. At night the sky in the Neutral Zone
glowed with a red light as the gas was flared or burned
off from the wells in the Burghan field.

Already the concession for the Neutral Zone exercised
by the Kuwaiti Government had gone to a consortium of
medium-sized American oil companies called Aminoil,
comprising Phillips Petroleum, Signal Oil and Gas, Ash-
land Oil, and a few smaller outfits, including Sunray,
with which Getty had wanted to merge his companies.
Aminoil had paid the Kuwaiti ruler the unheard-of price
of $7,250,000 down, with a guarantee of $625,000 in
royalties annually against 33½ cents for every barrel
of oil produced. Obviously, Aminoil also wanted the
concession for the Saudi half interest of the Neutral Zone,
which was available only because ARAMCO had given it
up in order to obtain more promising leases running
directly offshore from the high-producing Dharan field.

Getty was determined to win the Saudi concession,

thus beginning a long-lasting rivalry and competition between himself and Aminoil. The American Independent Oil Group, to give it its full name, was led by Ralph E. Davies, formerly the lawyer of Standard Oil of California, who had served as deputy petroleum coordinator during the Second World War, and who personally had an 8 per cent stake in Aminoil. He was soon to become one of Getty's most irritating *bêtes noires*.

Getty was so sure that oil would be found in the Neutral Zone that, even before he had the area surveyed, he offered King Abdul Aziz of Saudi Arabia an $8 million down payment for the Saudi concession, sight unseen. His next step was to choose a young geologist with past experience in Saudi Arabia to represent him. Paul Walton was a thirty-two-year-old petroleum geologist with Pacific Western who had a PhD from the Massachusetts Institute of Technology. He told Getty that the odds of finding oil in the Neutral Zone were 50–50, good odds in his opinion, compared with the USA where the success ratio was more like 1 in 10 to hit pay dirt at all, and maybe 1 in 40 to find an 'elephant'.

Sitting in the co-pilot's seat of a two-engine plane flying over the inhospitable desert of the Neutral Zone in the autumn of 1948, Walton picked out the only interruption in the terrain's monotonous flatness, two modest jebels some 50 to 100 feet in height that resembled upturned bowls with dark brown crests. The structures reminded him of a similar dome some 19 miles to the north in the Burghan field. He sent Getty a note in a plain envelope, announcing that he had sighted an anticline, a place in the earth's crust where the rock strata arched and then dipped, pushing the oil up from deep within the bowels of the earth.

Surprisingly, Getty wanted the young man to handle the bidding as well as the survey and had briefed him in the Pierre Hotel on how to proceed with the expected trying negotiations with the Saudi Finance Minister, Sheikh Abdullah Suleiman. Walton found Getty to have a half-mad look on his face. By the time Walton had finished the tortured bidding process, the stakes for oil production in the Middle East and the treasuries of the Arab kingdoms were never the same.

Getty won the concession, but only after making what seemed an unbelievable offer to the Arabs. He agreed to pay King Abdul Aziz immediately $9.5 million in cash to be used to pay the salaries of his civil servants. In addition, each year he would give the Saudis a flat $1 million minimum payment even if no oil was found, a promise that he tried to get out of when oil was taking longer to discover than expected. But what must have shocked and stunned the oil establishment in London, New York and Dallas was the promise to pay the Saudis the astronomical fee of 55 cents for each barrel of crude oil that could be produced. This was some two thirds more than the royalty Aminoil promised the sheikhs in Kuwait and, incredibly, two and a half times the 22 cents a barrel being paid by ARAMCO and the other Seven Sisters.

Paying the highest oil royalty in history was totally out of character. Perhaps Getty was making up for his rejection by the Iraqis in 1932; perhaps he was taking a long-term view about his investment. 'I told Arab friends of mine that I now felt I never had to run down an alley any time I saw an Arab approaching because I gave the Arabs very good terms right from the start,' he said later. He could very well have understood that exploitation of the primitive Arab sheikhdoms by the Western oil companies was coming to an end. After all, Venezuela, which was far closer to his ordinary base of operations, had won its fight to split oil revenues down the middle, 50–50, with the oil companies. Perhaps this was the shape of the future, and with oil at $2 a barrel a royalty of 55 cents was better than a 50–50 deal. Perhaps ARAMCO and the British would be forced to pay $1 a barrel royalty for their oil. Getty was prepared to pay the highest price because he knew it would not be the highest price for long. 'He knew the oil would be worth far more than $2 some day, if only he had the patience to wait,' says Garth Young, the veteran Signal Oil engineer who dealt with Getty over the Neutral Zone.

Even at $2 a barrel and a royalty of 55 cents, Getty could still make money. The cost of producing one barrel of oil in Saudi Arabia was about 20 cents, and this could be reduced, if the flow was great enough, to as little as 3 cents a barrel.

In order to get the concession, Getty even agreed to all the Saudi demands for a social infrastructure, including housing for his oilfield workers, schools for their children, formal training programmes for the Arab employees and even a small mosque for their religious needs. He backed down only when Suleiman demanded that he pay for the training of Saudi armed forces to protect the area against the possible incursion of Russian troops into the oilfield, a distinct possibility in those days as it is even today. Walton managed to arrange for Secretary of State Dean Acheson to send a wireless message explaining that no American corporation could legally pay for another nation's military forces.

Walton also had to cope with Getty's farfetched fear that the site of the domes he had surveyed, called Wafra, which means 'plenty' or 'abundance' in Arabic, was in effect 'an extinct volcano'. Walton assured him 'there was no sign of lava runs in the sedimentary rocks'. He was not about to buy Vesuvius.

After his successful negotiations, Walton returned

home to the United States with amoebic dysentery and parted company with Getty. He did not even receive a letter of thanks – merely a bonus of $1200.

The official assurance that Getty had won the concession was announced on 31 December 1948. What was not announced was that the wily Saudi Finance Minister, Suleiman, had secretly offered Aminoil the concession if they would equal Getty's offer. Luckily for Getty, Aminoil could not give the Saudi king more than they were giving the Kuwaiti sheikh without creating friction. Moreover, as Getty was about to find out, Aminoil was a consortium with several owners and voices on its board, all wanting their say, and it was almost impossible for them to come to a clean, quick decision about anything. Getty was answerable only to Getty. All the same, he would not have all things his own way. Like it or lump it, Aminoil was his partner in the enterprise, and the exploitation of the Neutral Zone was to prove one of the most difficult joint ventures in business history. Getty and Davies, the chairman of Aminoil, were egocentric, volatile types who could not share authority or agree on compromises.

There were to be furious fights about money. Getty had an obsession about cheapness and holding overheads down, and his passion for keeping expenses to a minimum was all the greater because Pacific Western was paying the Saudis far more than Aminoil was handing over to the Kuwaitis. 'Paul argued furiously about paying his share of every 100 feet of drilling pipe, and issued instructions to his field personnel that clashed with ours,' says Don Carlos Dunaway, the first president of Aminoil.

Getty was no more popular with the other oil companies in the Middle East than he was with his new partners, but he was quite a hit with the Saudis. The Lebanese lawyer representing Aminoil, Fawzi el Hos, told the US

Ambassador to Saudi Arabia, J. Rives Childs, that Getty's generous offer 'might well induce us to give considerable reconsideration to the basis of oil concessions generally in Saudi Arabia.' J. Skliros, general manager of the Iraq Petroleum Corporation, called the 55 cent royalty 'completely insane, uncalled for, and responsible for the difficulties being encountered in Iran and Iraq in renegotiating contracts.'

The Saudi Government was 'flabbergasted by the liberality of the Pacific Western Oil Company offer', according to Childs. Incredibly, he reported, Getty's winning offer became the 'impetus for the establishment by the government of a Ministry of Oil Development.' Getty, the first independent American oilman to win a concession in Saudi Arabia, was about to create the powerful ministerial position that would be held one day by Sheikh Zaki Yamani, the current Saudi Oil Minister.

It was a period of intense bitterness and acrimony between Pacific Western and Aminoil, between Getty and his opposite number, Ralph Davies. In the first place, Aminoil, the self-appointed operator of the joint concession, wanted to concentrate on finding the extension to the Burghan field coming down from the north at a depth of 3500 feet. But inexplicably the Aminoil operators chose not to drill on the crest of the physical structure, but off to one side.

Curiously, Getty had no desire to go to the Neutral Zone himself, so he sent his eldest son, George II, whom he barely knew. George, after being ousted as trustee of the Sarah Getty Trust, had served an apprenticeship as an independent oilman. At the age of twenty-five, young Getty became Pacific Western's Neutral Zone representative. He must have been astute, for he pointed out that Aminoil was drilling 'too low on the structure to produce', but for some unknown reason this mistake was not rectified for another three years.

Getty knew there was poor-quality crude oil to be found at quite shallow depths and wanted to drill dozens of cheap, shallow holes 800–1000 feet into the desert in order to bring up the vast reservoir of heavy-sulphur oil which the tests showed was there in abundance. Understandably Aminoil did not share his confidence that this 'garbage' oil could be turned into a commercial product because it was so hard to refine.

In Pacific Western's 1950 annual report Getty made plain his irritation. Since Pacific Western had borrowed $15 million to buy the concession, he decided to conserve his cash flow and stop paying a cash dividend. In early 1951 he reported confidentially to the Saudi king that 'we face the uncertain future with confidence that the operations in the Neutral Zone are being carried on in the best possible manner, and that if there is oil in the area, it will be discovered through these operations.' By mid-1951, as he approached the age of sixty, his diary became morose and self-deprecating about his financial acumen. He wondered why he was still active in business, but other entries reveal his ambition to become one of the richest if not the richest man in America. He was weighing up the possibility of building tankers in French shipyards and estimating the chances of selling oil to Greece.

Nineteen fifty-two came and went without any major discoveries in the Neutral Zone, creating even more tension between Getty and the Aminoil officials. Davies, as egocentric as Getty, refused to meet him after their first encounter when Getty glared at him with a concentrated angry stare for several minutes. Davies fled and was too intimidated to see Getty thereafter. Confidential dispatches from US diplomatic officials reported the enmity. Aminoil officials held 'strong personal and business-directed animosity to J. Paul Getty, making cooperation almost impossible,' wrote Harrison Symmes, the American consul in Kuwait. By early 1953 Symmes was telling Washington that 'company relations have evidently deteriorated to a degree that is alarming when their implications are considered.'

And then, after four years of frustration, success came suddenly. On 10 February 1953 Getty began drilling his sixth exploration well near the apex of the dome for the first time. This was presumably the location which George Getty II had wanted to explore in the first place. After an investment of $30 million by both partners, the Burghan producing sands were hit at a depth of 3482 feet. The discovery was estimated to be 250 feet in width, which indicated that Kuwait's Burghan structure, which had an oil column 873 feet wide, gradually petered out as it extended southwards. Symmes reported to the State Department that, 'barring unforeseen major structural unconformities, all evidence indicates that the Wafra strike is of major importance, even by prolific Middle East standards.' *Fortune* magazine declared that it 'is somewhere between colossal and history-making'.

The discovery of oil in the Neutral Zone had a wondrous effect on Pacific Western stock. In the space of one month, March 1953, it doubled from $23.75 a share to $47.75 a share, causing Getty's wealth to multiply by precisely the same factor. At the time Getty felt the stock was worth at least $85 a share in a sellout or liquidation.

In early April 1953, *Time* magazine reported the Neutral Zone success and featured James MacPherson, a Scot who had moved over to Aminoil as general manager from ARAMCO, along with Ralph Davies. Getty was barely mentioned.

Rather than make peace, the oil strike only intensified

the quarrels between the two sides. Davies went through the roof when he heard that Getty had met the Kuwaiti ruler, Sheikh Abdullah bin Salim al Sabah, privately in Switzerland to suggest a joint operating company in the Neutral Zone which would give Getty more leverage. According to the American consul in Kuwait in a report to the State Department, 'it is dog-in-the-manger for the Pacific Western President to suggest to the Ruler of Kuwait that the Ruler exact similar terms from Aminoil. That could only complicate an already almost impossible situation.'

Garth Young who represented Signal Oil and Gas in the Aminoil set-up, found it 'a miracle that oil was produced at all in the Neutral Zone under two different sovereigns, two separate tax and custom structures, two different concessions, and warring personalities.' According to him, Getty's determination was the crux of the matter. 'Getty wanted the oil. He knew it had a lot of value. He could see further than any other guy I knew. Oil to him was the valuable thing, and it didn't matter that he was paying 50 per cent more than anyone else.'

During the whole of this period Getty never once visited the site. Incredibly, it was not until 1954, a year after the first oil was discovered, that he went personally to the Middle East for the first time in his life. (Davies had such a 'pathological fear of dust, dirt and germs', according to Garth Young, that he never went to the Middle East at all.)

Getty left Venice on the Orient Express with a tall, strapping American lawyer, John Pochna, who had been to Saudi Arabia and who was to be his eyes and ears there at the then stupendous salary of $75,000 a year. He had no contract, however, because Getty wanted to be able to get rid of him quickly if he became *persona non grata* with the Saudi royal family. They broke their journey in Istanbul to visit the museums and historical sights. On the train between Smyrna and Baghdad, Getty told Pochna he wanted to leave money 'to every single one of the 100 lovers' he had had by the age of sixty-two. He made up a list and told Pochna to handle it, but nothing more came of the talk.

On arrival in Kuwait, Getty met MacPherson of Aminoil, C. J. Pelly, the British political agent in Kuwait, and Symmes, the American consul. Symmes thought him an 'aberration – he seemed so remote. His hair was dyed red in an obvious fashion and his facelift was visible to the naked eye.' Still, he impressed the American official because he was not 'typical of the arrogant rich man'. His 'softly, softly' approach, as Getty himself described it, made him appear vulnerable.

Although Getty liked to play by the rules, he could not accept the result of a 1954 Saudi law establishing a tanker company owned by Aristotle Onassis which was given a monopoly to transport crude oil from Saudi Arabia. To obtain the law, Onassis had bribed important Saudi officials, a flagrant affair, which infuriated not only ARAMCO but the State Department, the CIA and Onassis's most bitter competitor, the Greek shipper Stavros Niarchos.

Getty began negotiating with Onassis in the summer of 1954 and agreed to lease several tankers from him, according to Pochna. Pochna says that Getty differed from Onassis in that he 'was more cerebral, more cautious than Onassis'. Getty did not like to tangle with foreign governments. Onassis, according to Pochna, 'thought he could do anything, could have his way by bribing the right people.' The two men became friends because they respected each other's business acumen and success with women. 'At the first Getty–Onassis meeting, Ari was complaining about the oil companies kicking the shit out of him and putting him in their doghouse,' says Pochna. 'Getty said to him, "When you went into the doghouse, did you happen to notice a pair of shoes there that were mine?"'

13

The Nomad

I have often maintained that I possess a rare talent and strong
inclination to be a beachcomber.

J. Paul Getty

To build his empire in the Middle East Getty decided to
locate himself in his favourite playground – the capitals
of Western Europe, halfway between the sands of the
Neutral Zone where he expected to find oil and southern
California where his underlings were required to carry
out the boring details of the day-to-day administration of
the company. It was meant to be a temporary arrangement. In fact, his nomadic existence lasted for eight years,
from mid-1951 to 1959, from his sixtieth to his sixty-
eighth year.

Such a life had always appealed to him. He would be
free of the constraints of married life and fatherhood.
His book-keepers could dole out the alimony and child
support, as they had done before during his absences.
His marriage to Teddy was never a total commitment,
and their son Timmy was in bad health. The boy had
suffered from eye trouble from birth and, in 1952, when
he was six years old, a tumour the size of an egg formed
between his eyes and 'affected his optical nerves and
eyesight', according to Ware Lynch, Teddy's brother.
Timmy would require several operations. According to
Getty's friend, Art Buchwald, then a columnist on the
Paris Herald Tribune, Getty 'had no use for any of the
kids. He never spoke of one with pride.'

As for his business empire, the telephone and telegraph

would keep him in touch. He told Dave Staples that he 'preferred Europe because he could travel with two or three women at the same time.' He moved from the Ritz in London to the George V in Paris to the Flora in Rome, ostensibly to develop markets for the oil he was producing in the Middle East, but also to acquire new lovers and to add to his art collection.

Getty had an extraordinary magnetism for women. 'Even if he was disagreeable, the women were always sniffing about,' recalls Paul Jr. It was probably the result of a unique combination of wealth, power, a well-informed and cultivated mind, impeccable manners, a wry sense of humour and, apparently, tremendous sexual prowess. The proof is that when he left the United States in 1951 there were many broken hearts. For years afterwards women friends sent him letters of distressed love, about the excitement and the pleasure he had given them, begging him to send them some word that he cared about them and pleading for the date of his return. Their devotion to him is impressive, if sad. 'Oh God, to see you once more,' wrote one. 'I am worried about your silence,' wrote another. A third, whom he had met in the thirties, was still writing to him twenty years later.

He was, of course, sending money to several lady friends. Even $100 at that time was worth a lot. The recipients included a lady in Los Angeles, another in Atlanta, a third in Mexico, a married lady in New York, and two girlfriends in California. He even arranged a rendezvous in Europe with a girl he had picked up at Santa Monica railway station in 1946, about the time that Timmy was born. Despite their apparent devotion, he often hired private detectives to investigate them and their backgrounds.

The wealthy expatriate added to his American gang of girlfriends a new group of lovers of all nationalities.

Although he was in his sixties, he could pass for a man several years younger. Erect and wiry, 5 feet 11 inches tall, he had the type of looks known as *joli-laid*, with an aura of danger and mystery that women found seductive. He also had a fluent knowledge of French, German and Italian. Some of his friends thought that he passed beautifully as either a Frenchman or a German. He kept an entire wardrobe in the basement of various European hotels so that he could assume the identity of the appropriate nationality. He liked to brag that his English accent disguised his American origins when he was at Oxford, although this claim sounds apocryphal.

He sometimes used the procurement services of a painter friend, who seemed to know the pretty debutantes available to travel with him for a few favours. Getty had met this man in a Berlin night club in the late 1920s at a time when he was in the diplomatic service. Getty's new friend was a short, intense raconteur who sported a moustache and had a taste for young girls. Together the two men lived a demimonde existence among café society. At one point he offered Getty M., 'an exceptionally intelligent and well-educated person' whose 'charm is mysteriously beautiful'. In a letter to Getty the painter suggest that Getty should 'be generous in every way', since M. does not know what a 'woman-rascal' Getty is.

The two men's friendship was a reflection of the bohemian strain in Getty. He was bohemian in the sense that he was easily bored by the constraints of a home and the company of a single woman. Moreover, none of his women friends seems to have been particularly interesting at an intellectual level; none had a college education or a profession, except for Teddy. His unconventional personal life would have been unacceptable to his mother, and the pressure to conform to the social norm was greater in America in the 1930s and 1950s than in Europe.

Whereas everyone in Los Angeles knew who Getty was, he could move about freely in Paris, London, Berlin and Rome. It was much easier to indulge his sexual proclivities on the Continent than in the conventional, church-dominated strata of American society in which his parents lived.

Getty received thousands of love letters but never wrote any himself. He never documents his love affairs in his diary, nor does he ever describe a single intimate situation. He never referred to a woman's anatomy in public – he was far too reserved and subtle for that. But an attractive, primitive drive kept him constantly on the go, even with his secretaries. Barbara Wallace, his secretary for the last twenty-four years of his life, who admired and respected him greatly, wryly recalls his offer to go beyond their customary relationship. 'We avoided that one,' she says fondly. 'One can always say no.' Another of his secretaries was not so sage. In the early fifties Getty employed a young, well-bred English girl to accompany him to Rome as his secretary. Once there, she met the youthful, dashing son, Paul Jr, and proceeded to have an affair with him. When Getty found out he sacked her.

Getty was not phlegmatic about the possibility of loss in his relationships with women. His divorces caused him 'loss of sleep, appetite and peace of mind,' he wrote to one of his lady friends. When 'darling Penelope' (Kitson) was on holiday at her home in Portugal one summer, he warned her sweetly: 'Do not go far from shore in your little boat; you never know when a wind might spring up suddenly and make it difficult or impossible to get back.'

Above all, in his new milieu he enjoyed the company of the most glamorous, wealthiest and most beautiful society that Europe could provide. He wined and dined with Aristotle Onassis, Baron H. H. Thyssen-Borne-misza, heir to the Thyssen steel fortune and owner of perhaps the greatest collection of old-master paintings in the world, the Duke and Duchess of Bedford, who owned Woburn Abbey, the Duchess of Argyll, an American beauty who had become titled by marriage to a Scottish duke. In Paris he was the guest of the charming fellow philanderer, Sir Charles Mendl, who had a marriage of convenience with the elegant interior designer, Elsie Mendl. In Rome, he often dined with Lord David Beattie and his wife Adele, the charming sister of the movie star and dancer, Fred Astaire. When he was in London, Getty often stayed at the flat of Lady Drogheda; she also allowed him to use her car. He was photographed in a Paris night club with Dorothy Spreckels, a flame from his New York days. Among his friends were the Woolworth millionairess Barbara Hutton and her movie star husband Cary Grant, known together as 'Cash and Cary'. Aly Khan was another.

In spite of his love of mixing with the rich and famous, Getty's own living conditions hardly measured up to the 'gracious and cosseted lives people lived in the eighteenth century', as he wrote about his favourite age in *Europe in the Eighteenth Century*. True, he had a suite in the fashionable George V Hotel in the 16th arrondissement of Paris, where he registered, mysteriously, as 'Monsieur Paul'. But suite 801 was hardly a palatial apartment. One visitor described its 'wild disarray . . . an unmade day bed, the cold remains of a meagre meal . . . It might have passed for a bookie's office or a convention caucus room.' He seemed to be living out of a suitcase and conducting his business out of shoeboxes. The suite was reached along a narrow passage partly blocked by pipes. In London he occupied the smallest and cheapest suite of rooms, 611–12, that the Ritz could provide. When he was not in residence at either hotel, the porter was instructed to forward his mail by re-addressing it, rather than enclosing it in new envelopes. This saved postage.

The stories of Getty's pathological parsimoniousness are legion. Even before he acquired international fame he used to complain that he was always expected to pick up the tab for dinner, although he seldom did. At the end of one dinner party given by the gossip columnist Elsa Maxwell at Maxim's in Paris, Getty was handed the bill for the crowd of smart international figures she had assembled there. He dug into his pockets, left only the price of his meal and walked out, leaving behind a stunned and fuming Elsa Maxwell.

One of Getty's oldest women friends always felt that Getty could not bear to think that he had so much money. She recalls one occasion when, after dining out, Getty was brought his change. He looked at the money and said, 'All that for me?' For her, 'Paul was as mean to himself about money and the good things in life as he was to others.'

The Simons, a European couple who saw Getty in Paris and the South of France, 'disliked him as a person' but found him 'extremely erudite'. They were expected to pay for dinner even though they 'were comparatively poorly off at the time'.

And Getty never invited his friend Frank Ryan to dinner in his entire life. They had first met at the Olympic Games in Berlin in 1936. In the 1950s Ryan was in Paris, as president of the World Commerce Corporation, an international trading concern which also performed an intelligence function for the US Government. He moved in the fastest sets and chased the models and actresses who gathered around the rich Americans. He called on Getty one day and, on behalf of his family, offered $17.5 million in cash for the Pierre Hotel. At that price Getty would have made a tidy profit of over seven times the $2.4 million he had paid for it in 1938. He thought a while. 'Why, yes, Frank,' he said, 'that sounds like a fair price. Why don't you cable my lawyer in New York and tell him about the deal?'

A bit nonplussed, Ryan suggested that Getty send the cable, since it was *his* hotel and *his* lawyer. Getty insisted. Ryan, in effect, had to pay for the cable, which Getty dictated, to his lawyer Dave Hecht in New York: 'J. Paul Getty agrees to sell the Pierre Hotel to Frank Ryan and his brothers for $17.5 million.' Getty hesitated for about three seconds, and then added: 'Unless you can do better in ninety days.' So, rather than make a firm deal, Getty used Ryan's unsolicited bid to prod Hecht to obtain an even higher price in New York, which he subsequently did.

Who can tell whether it was his business acumen, cultural aspirations or sexual drive, or a combination of all three, that mesmerized such a diversity of women during these years, women of every nationality and from every social milieu? He met Ethel LeVane, a British-Jewish art dealer and writer from Manchester, on the Super Chief express train from Chicago to Los Angeles. (His excursions on transcontinental railways and transatlantic voyages often provided opportunities for seduction.) She was petite, attractive, but hardly beautiful, and Getty claimed he made love to her twice in his private compartment on the express. In the 1950s, when he was in Europe, they travelled widely together collecting new works of art, visiting museums and investigating the origins of his earlier major acquisitions. They even collaborated on an art book, *Collector's Choice*. Getty contributed three short but imaginative chapters at the end on life in ancient times. The earlier chapters by Ethel LeVane are about an art expedition by Getty through

Europe, accompanied by an imaginary, short, pot-bellied, balding but smart Dutch art dealer called Mueller. Mueller is, of course, a cover for Ethel LeVane. Her description of Getty is most affectionate and flattering. Beneath the stern, forbidding coldness there is an appealing boyish quality.

The art dealer saw in Getty a reflection of the father: an astute business brain, an agile though conservative mind, an insatiable thirst for knowledge – which he appeared to absorb with the minimum of effort – and a reticence which made knowing him difficult. He wondered whether anyone really knew this complex individual, whose personality embodied so many and such varied facets.

He was so complex and reticent that it was impossible to know him completely. Mueller 'admired his restraint, his unpretentious dignity, his ability to absorb information with a minimum of effort.' When he was called away from this art odyssey with LeVane, he said it was because of 'moral responsibility', which meant either that he had to take care of the Getty oil interests or that he had another woman parked somewhere.

Mark Goulden, of the London publisher W. H. Allen, became a friend of Getty through the publication of *Collector's Choice*. He was 'reasonably sure he had a real affection for this garrulous, lively little woman who was so obsessed with Getty that she talked (interminably) about little else.' She washed his socks and underpants, stayed in the clinic during his periodic facelifts, and (so Goulden alleged) introduced him to 'many a nubile girl until he began to suffer from impotence, to counter which he took pills that seriously affected his health.' Dinner guests at Goulden's West End flat would hear her talking babytalk on the phone to her lover in Paris. 'Is my Pauly-Wauly happy in naughty-waughty Paris?' she was overheard asking.

Towards the end of April 1953 Ethel LeVane accompanied Getty on a trip to Florence, where he bought four oil paintings including A *Lady Playing the Guitar* by Bartolomeo Veneto. While there, they visited the American art connoisseur Bernard Berenson, who lived at the Villa i Tatti. Getty's impression, as recorded in his diary, of this most fascinating character in the world of art is banal in the extreme, confined entirely to his appearance: 'Always one of the best-dressed men. About 5' 2" tall, he weighs around 120 pounds. He walks with a cane. Although he is now 88, he looks a vigorous 70 or 75.' That was all. For his part, Berenson was more forhcoming about Getty whom, he recorded in his diary, he had met 'through an accidental encounter in a hotel . . . I took a mild interest that pleased him, and it led to his sending me photos of Greek fragments I approved of, and then to his coming to spend a few days here. I got to know more and more about him, chiefly through his collaborator, but not him.' A few days later Berenson was told that 'Getty's ambition was to be the richest man on earth', and that he nearly was.

Getty must have been thrilled at Berenson's praise of *Collector's Choice* and his advice that Getty should forsake his business career for writing. Berenson opined that a switch to writing would make Getty a much happier man, which was exactly the notion Getty sometimes had himself. In fact, he was a terrible writer. *Europe in the Eighteenth Century* is a stilted, banal compendium of facts. Nor could he have been a diplomat because of his inability to travel by air or by sea. He did not have the personality to be a diplomat because of his need to be in absolute control of any venture.

After Italy, Getty and LeVane headed for England. Lord Lansdowne, owner of a collection of superb classical statues, had just died and Getty, who was interested in

one particular piece, thought he might pick it up cheap.
He had his eye on a marble statue of Hercules which had
been found in 1790 in the grounds of what had been the
villa of Getty's hero, Hadrian. He sometimes told friends
that he thought that he was the reincarnation of the
Roman Emperor. Although he found the face of Hercules
weak in character, a 'work of decadence', the fact that it
had belonged to Hadrian was more than enough to whet
his appetite. The estate of Lord Lansdowne had placed a
reserve price of $40,000 on the statue. Getty put in a bid
of $24,000. To everyone's surprise, it was accepted.

Doing the rounds of the galleries and the art centres of
Europe left Getty with plenty of time and opportunity for
socializing. In 1954 he visited the London art gallery of
Sir Robert Adby. On this occasion he did not buy any
paintings, but he did meet Sir Robert's tall, slender
assistant, Penelope Kitson. Six feet in height, and a
prim, almost prudish, dresser, Penelope was married,
unhappily, to a Cornish landowner, Robert Kitson. They
had three children. An earlier marriage during the war to
a British naval officer had also failed.

Penelope joined the Getty payroll. The Getty oil
empire, having discovered oil in the Neutral Zone, was
now building supertankers. On each vessel Getty built an
owner's suite. On the first ship, the *George F. Getty*, he
disapproved of the decorations and asked Penelope
Kitson to be responsible for the next two, the *Minnehoma
Getty* and the *Oklahoma Getty*. The suites were lavish. In
each there were two bedrooms, two bathrooms and a
luxurious sitting room. Strangely but perhaps typically,
Getty never used these costly owner's suites.

While he was living it up in Europe, his youngest child
Timmy was undergoing painful brain surgery in Los
Angeles. The tumour, pressing on the optic nerve, was
making the child blind, but Getty never visited this son

whom he professed to love more than the others. He kept Teddy on a tight budget and constantly complained about the size of Timmy's hospital bills. Against all the odds, the surgery appeared to be a success and the boy seemed to be recovering. So, in 1955, after Getty's four-year absence, Teddy took her son to see his father in London and Paris.

Despite his disfiguring illness, Timmy was a red-haired, freckled Huckleberry Finn. He must have been puzzled by his father's relationship with other women, for he told his mother that Penelope did not know how to make a peanut butter sandwich. For her part, Teddy was amazed at the way Penelope dominated Paul.

If there had ever been any hope of a reconciliation between Teddy and Getty, it was killed on that trip to London and Paris. Teddy had hoped that he might come back to America but he refused. He did, however, ask her to stay in Europe 'and I will make you as rich as Queen Elizabeth'. Teddy refused. She left with Timmy for America after only a short stay in Europe. She was probably aware by now that Getty was unlikely ever to return and that her multimillionaire gypsy husband cared more about money than he did about the most priceless thing in the world to her, her sick son Timmy. He had no emotional ties with any of his ex-wives or their sons, or with Teddy or Timmy, regardless of what he professed in his diary. The thought of divorce was already looming in her mind.

Getty constantly wrote in his diary about returning to sunny California but, like so much of his other behaviour, the diary entries seemed only to reflect what he knew others expected of him. Or was he blocked emotionally from fulfilling his own wishes? It would have taken a psychiatrist years to find out the true answer, and then the likelihood is that there were several answers.

His inability to return home, like many aspects of his life, gathered its own mythology. Acquaintances tell the story that he consulted a fortune-teller in Europe, who told him that if he crossed the Atlantic one more time, either by sea or by air, he would die. His close friend, the Duchess of Argyll, who also believes in fortune-tellers, gives credence to this tale. When the Italian liner *Andrea Doria* sank in 1956 in Nantucket Sound after a collision, it was said that Getty had booked a ticket on the boat but had cancelled at the last minute. From that day onwards it was always the same pattern of events. He would make a reservation to sail back to the United States but would always cancel, claiming pressure of work.

In any case, he could run his business from anywhere in the world. He needed no corporate headquarters, no committee meetings, no swarms of aides. He carried boxes and files of business papers with him wherever he went, piling them high in hotel rooms, whether it was the Ritz in London or the Hôtel George V in Paris. And this way of life was so helpful in satisfying his insatiable appetite for women. He slept with a clutch of American and European women who chased after him and who took his fancy. But they were not wined, dined and bedded in multimillionaire fashion. His girlfriends were not provided with suites adjoining his own simple suite at the Ritz in London; they were unceremoniously dumped in the former maids' rooms under the eaves of the hotel.

Sometimes he did not get the ladies into bed. He chased assiduously after the titled widow and German industrial consultant, Baroness Marianne von Alvensleben, from the moment he met her in Baden-Baden in the early 1950s. 'He was in love with me, but he was not my cup of tea. It's very simple. He was very depressed and couldn't understand it. He called me every day and asked me questions about business and advised me not to work so hard.' Yet their friendship remained close until Getty's death. She found him 'courtly, with perfect manners and so polite'.

He always claimed that he could not enjoy any close male friendships because men were jealous of his wealth. In reality, he was afraid of being taken for a ride. However, he was able to form a warm attachment for one man at this time, a charming and wealthy Frenchman, who needed nothing from him and therefore could be trusted. Rather the reverse. Getty could learn from Paul Louis Weiller, whose love of beautiful women and works of art, probably in that order, equalled that of Getty.

The son of a wealthy Alsatian member of the French Parliament who owned a large aircraft company, Weiller had been a fighter pilot hero during the First World War, and had been rewarded with his own postage stamp and the title of 'Commandant' which he revelled in. He was a slim, dark and exotic French Disraeli, who owned several beautiful homes in France and seemed to be an eternally youthful *bon vivant*. Although he was often reviled by some sections of French society as 'that rich Jew' (he owned a chain of petrol stations in Germany), Weiller was on good terms with Lord Louis Mountbatten, the British royal family and everyone who was anyone in European society. He shunned publicity, while using his enormous wealth to support the arts.

Weiller found Getty to be *un homme de Balzac*, a man from the countryside, a hick who did not have the sophistication of Weiller's friends. Getty advised him to promise no guarantees on the sale of his 420 petrol stations, 'a wonderful piece of advice which helped us make a great deal,' says Weiller. In return, Weiller taught Getty a knowledge of high society.

Each August Weiller held open house at his summer villa, La Reine Jeanne, near Le Lavandou on the French

Riviera, to which all the beautiful people were invited. He enjoyed a combination of stimulation and beauty. Often the Hollywood set, including the Charles Chaplins and the David Nivens, would rub shoulders with royalty such as Queen Soraya, wife of the Shah of Iran, or with politically influential intellectuals such as André Malraux, or industrialists like Aristotle Onassis. So valued were his taste and judgement that Onassis brought his prospective fiancée, Tina Goulandris, to Weiller for his personal verdict on whether she was the right woman for him to wed. Claus von Bulow, then a handsome young barrister who was to work for Getty and whose own marital troubles were to hit the headlines many years later, was another visitor at La Reine Jeanne where, he said, the young beauties were so plentiful that 'it was impossible not to get laid'.

Paul Louis Weiller always felt he was Getty's one true friend. He was too rich to need any favours from the American oilman, except perhaps the latest information on how well Getty Oil was doing. To Getty, Paul Louis was a 'life-enhancer', 'at once the shrewdest of business-men and the kindest and most considerate of human beings, a man of impeccable taste and manners, yet without the slightest trace of prejudice or snobbery.' Although he lived a far more sybaritic existence than Getty, he maintained the lowest of public profiles and always advised his American friend to do the same. 'I told him if he opened his home to everyone, the *sans culottes* [revolutionary rabble] would get him, but he never listened,' Weiller says.

It was through Paul Louis Weiller that Getty met a woman who was to play a provocative role throughout the rest of his life. Sooner or later, even without Weiller, Getty would probably have met the temperamental Russian aristocratic beauty Mary Teissier, with whom he was to have a torrid affair and come within a hair's breadth of marrying. From the moment he met her at Weiller's small charming town house in Neuilly on the outskirts of Paris, Getty fell under her spell. She was a hauntingly thin beauty, who claimed blood relationship with the Romanovs. Her marriage to an extraordinarily handsome Frenchman, Lucien Teissier, had fallen apart because of his amorous exploits. The Duchess of Bedford once saw a radiant Mary enter a Paris restaurant with light snow on her fur cap and on the two borzoi dogs she was leading. 'It was the most beautiful aesthetic sight I ever saw in my life.'

Getty was so bowled over by her beauty, her lively nature and her aristocratic blood that he wanted her to obtain a divorce. She told everyone that he was a demon lover. He was ready to make her his sixth wife when, according to Paul Louis Weiller, she made '*le baiser de la mort*', the kiss of death. She began to talk about the money Getty ought to give her. '*Elle est une Russe,*' says Paul Louis. 'She is a Russian, who kills the things she loves.'

Meanwhile, Getty's fifth wife, Teddy, had finally seen the writing on the wall, and in 1956 she filed for divorce in Los Angeles. The following year Timmy's growth reappeared and the boy underwent three terrible operations. The divorce became absolute on 29 May 1958. In the August of that year, Timmy, seemingly cured, went into hospital for an operation to reset the bones of his skull. The boy, who had known little but physical pain during his short life, died under the anaesthetic. According to Ware Lynch, 'When he died, Timmy was blind and had a misshapen head. I called Paul to announce his death. He had to attend to business,' says Lynch. And 'Teddy paid the bill at the Pierre Hotel.'

In fact, Getty learned of his son's death while visiting

his friend Heinie Thyssen at La Favorita, a splendid villa facing Lake Lugano in the foothills of the Italian Alps. There Thyssen, another 'life-enhancer', kept perhaps the most magnificent collection of old-master paintings in private hands in the world. It never entered Getty's head to return home. He wrote in his diary, as if to excuse himself in the eyes of those who might one day read it: 'Had I not been assured the operation was a slight one I would have gone to New York to be present but there was no urgency about the operation. Dear Teddy! How brave she is! Darling Timmy! The world is poorer for your loss and I am desolate.'

Apparently Getty could feel emotion, but he didn't want anyone to know about it. It was a sign of weakness, not strength. 'He was afraid of emotion, terrified of emotion,' according to Gail Getty, then Paul Jr's wife. When his close friend Jack Forrester died in 1959, the year after Timmy's death, Getty called Gail and told her, '"Please, I can't go to the funeral. Would you go for me?" He was devastated.' He used his diary to set the record crooked, not straight, and in it he practised his own private brand of self-delusion and hypocrisy. Ironically, he treated his dog Shaun better than his son. When the dog was suffering from a tumour, Getty had the best veterinary surgeon flown in and spared no expense. When the dog died, he stayed in his room for three days weeping.

Getty was far too busy in Europe to return for Timmy's funeral. He was negotiating to buy an Italian refinery, and business came first. In his autobiography the death of his son is recorded in passing in one and a half lines, wedged between a passage on what he saw as a dangerous political upheaval in France and a report on his business activities.

Now there was even less reason to return to the United

States. The opportunity for a rich bohemian oilman to be on a continual grand tour, soaking up the culture and the hedonistic lifestyle of the Continent, proved to be a strong opiate. Frederico Zeri, one of his close art historian friends, felt 'Paul stayed in Europe because the Old World seemed safer and more harmonious than the New World.'

14
The Eighth Sister

If you can count your money, you don't have a billion dollars.
J. Paul Getty

Despite his nomadic existence and diverse private entanglements, Getty was able to concentrate on the complex affairs of his business life and create the enormous wealth which made him the richest man in the world by the end of the 1950s. He had a goal and a programme of how to get there. He would mine the cheap, shallow crude, the 'garbage oil' of the Neutral Zone. He would transport it to the United States and elsewhere in his own tankers. And, when it arrived, he would refine it in his own refineries.

In order to build this colossal integrated enterprise he totally changed character, in so far as he was prepared to invest $600 million in the project, a sum 200 times the size of the original Sarah Getty Trust. If his mother had been alive, she would have thought her son had become exceedingly reckless again, despite the massive expansion of the American economy during the 1950s.

He was taking on the Seven Sister's monopoly of world oil markets, but he had the advantage of being a lone wolf, sole proprietor of his own empire. To raise the money he required he decided to use the credit of Tide Water, or Tidewater as it became once Getty owned an overall majority of the shares.

He achieved this with a series of typically shrewd financial moves. Since he was using Pacific Western's borrowing powers to finance the Neutral Zone exploration, Getty had to find other resources to enable him to continue buying Tide Water shares. He arranged for Mission Corporation to borrow $14 million from the Chase National Bank and used that money to buy more Tide Water. His long-range plan was to supply roughly half Tide Water's crude oil requirements from his own Neutral Zone production. Even before any oil was produced, Getty ingeniously reduced the number of Tide Water shares so that he would own a majority. He offered the unsuspecting shareholders a preferred stock that paid a small cash dividend instead of the common stock they already owned which at the time paid no cash dividend at all. Most shareholders prefer to accept a regular income rather than take the long-term view that the owner has some plans that might make the common stock far more valuable in the future. Leon Levy, now a mighty Wall Street potentate, but at that time a young analyst at Oppenheimer & Co., recognized intuitively that Getty was not likely to stop at 51 per cent ownership. He felt that this scheme to obtain voting control of Tide Water foreshadowed far greater investment profits in the future.

So it was to prove. With voting control of what was now the largest enterprise in his empire, Getty was able to grow by leaps and bounds through his daring expansionary programme. He would invest $200 million in a new refinery on the east coast of America that could turn the crude oil from the Neutral Zone into gasoline for automobiles. Another $60 million would be put into the aged refinery in Avon, California, that desperately needed an overhaul. The number of gasoline stations flying the Tidewater colours would be doubled at a cost of $120 million as outlets for the gasoline being produced in Delaware and Avon, California. He would spend $207 million to build the first supertankers to carry the crude oil to Europe, America and Japan. 'My competitors have been burdened with small tankers,' he told *Time*

magazine. 'I am determined to obsolete their fleets by
building no tanker less than 46,500 tons.'

The $200 million refinery was built 15 miles south of
Wilmington, Delaware, and was to provide a total of
130,000 barrels of Flying A and Veedol products for
Tidewater's chain of gasoline stations. It was a construction project worthy of Hadrian and the wall he built
across the north of England to keep out the Picts and the
Scots; some 3 million cubic yards of earth were moved
and 800 acres of marshland were raised 12 feet above sea
level, creating a 40-foot-deep channel up the Delaware
River to give Getty's tankers access to the port. Far
smaller amounts were spent to build a refinery in Denmark and to acquire one in southern Italy as a way of
gaining entry to the northern and southern gateways to
Europe. It was obvious that cheap oil was going to replace
coal as Europe's chief source of energy in the 1950s;
between 1950 and 1965 oil's share in providing energy for
the Common Market nations rose from 10 to 45 per cent
while the use of coal fell from 74 to 38 per cent.

Across the Atlantic, Tidewater could use Middle Eastern
crude in great volume. It was far cheaper to import it than
to extract it inside the continental USA. Shipments began
in September 1953 and in two short years Pacific Western's
fifteen producing wells in the Neutral Zone were pumping
4.4 million barrels of oil, more than that produced by its
556 wells in the USA. By 1957 Tidewater was purchasing a
large portion of Pacific Western's Middle Eastern crude
from the Neutral Zone, for a cost of $21,874, 306 and by
the end of the 1950s had doubled the number of its service
stations to 3885. The company now ranked seventh in its
share of the total US gasoline market.

Getty only narrowly missed another major coup in
1955 – a 10 per cent participation in Iranian oil production
which was being divided up among the independent

American oil companies after the Shah was restored to power in Iran. The State Department, through the offices of Price Waterhouse, the international accounting firm, awarded Pacific Western and Tidewater 5 per cent participation each in that part of Iranian output. The political power of the Seven Sisters, however, reduced Getty's share, so that he got less than 1 per cent of Iranian oil production, a kiss rather than a wholehearted embrace. In any case, Getty received about $1 million a year in revenues without having to invest a nickel.

He also missed out on the most productive area of the Neutral Zone, the offshore area, where a Japanese group obtained the concession and, late in the 1950s, began producing a much higher-quality crude oil than that found onshore.

Nevertheless, Getty had achieved his ambition. He had obtained a foothold in the Middle East. On 21 April 1956, at the age of sixty-four, putting retirement behind him in a symbolic gesture, he changed the name of Pacific Western, which had little significance for the Middle Eastern crude it was producing, to the Getty Oil Company. For the first time since the 1930s the family name was placed at the top of the pyramid, reflecting his decision only to 'retire when they put me six feet under'. He was determined that the most valuable asset in the Neutral Zone, the 'garbage' oil to be found in shallow wells, which Aminoil did not want to produce, should make him a fortune. It was of much lower quality than the oil coming from the deeper Burghan sands, but there was more of it, much more, and Getty meant to bring it up, turn it into a commercial product and find a market for it. It was only saleable at a discount well below the market price for the best product, say at some level below $1.60. 'Maybe the crude oil was garbage,' says Everette Skarda, one of Getty's wisest field engineers in the

Neutral Zone, where he worked for many years, 'but the cost of lifting it was so incredibly cheap. Paul loved drilling 1000 feet into the shallow Eocene zone for a mere $25,000, far less than a well in California. Often it cost only 11 cents a barrel to bring it up if the pumping units were used. Then there was a 6 to 8 cents a barrel cost to process the oil in a small nearby refinery. The overhead was kept at rock bottom, which often meant skimping on personnel and subsistence. Paul never allowed more than 700 employees in the Neutral Zone.'

With such penny-pinching, Getty was able to produce Neutral Zone oil for 20 cents a barrel, on top of which he owed another 55 cents a barrel royalty to the Saudis. This made his overall costs 75 cents a barrel before transportation. Any oil he could sell on the world market at a price of over $1 a barrel represented a good profit for Getty Oil, to the chagrin of Aminoil. Their treasurer, Joe Cumberland, saw him as an enemy not a friend, but even he had grudging praise for him. 'He could tell you every single well, where the casing was set, what it cost to drill, what it was producing. He had it all in his head, an intricate knowledge of geophysics, engineering. It was really impressive.'

Intimately as he knew the details of all the wells, he wanted to make sure that the petroleum engineers knew them equally well. He would play little tricks to check on the extent of their knowledge. Once, when Garth Young visited him at the Ritz, he noticed that Getty had changed some of the technical data on the wells. This was a ruse to see if the engineers knew what they were talking about.

In 1958 Getty's petroleum engineers, DeGolyer and McNaughton of Dallas, Texas, celebrated his tenth anniversary in the Middle East with a splendid present – an estimate of oil reserves in the Neutral Zone of over 12 billion barrels, and most of it, 75 per cent to be exact, was recoverable from Getty's own shallow wells. It was sour and corrosive, but he knew how to sell it to the world. Aminoil had no captive markets and 'they could barely give the stuff away,' says Skarda.

Back in his days in Oklahoma Getty had learned that the frenzied exploitation of an oilfield meant that the underground reservoirs produced much less oil than they would if the drilling and extraction rate were carefully controlled. He recognized that each well in the Neutral Zone could produce 50 per cent more oil than DeGolyer and McNaughton estimated, and the extra 50 per cent could be turned into profit. He treated each well in the Neutral Zone as a separate commercial enterprise, as its own profit centre, attributing expenses to each well separately. In a short span of time he recovered every dollar spent in the Neutral Zone.

To build his tanker fleet, Getty again made the greatest use of every advantage that came his way. His friend in Paris, Paul Louis Weiller, arranged for him to build the supertankers in French shipyards, where this kind of work was subsidized by the French Government. 'Paul Reynaud, a Member of Parliament from Dunkirk, came to me and said, "We have no orders in our shipyards. Maybe your friend Paul Getty could build some tankers,"' says Weiller. 'And I said to Paul, "It's a good thing for you to be a friend of France."'

A key role in the French tanker-building programme was played by another of Getty's close friends in Paris, Jack Forrester, an American expatriate, a former OSS agent and a man of great charm who enjoyed womanizing with Getty. Art Buchwald, who knew both men and observed their goings-on from the sidelines, found Forrester 'a really interesting guy, the only guy who wanted to spend time with Getty.' He had no regular job, but he was well connected. It was Forrester who conducted all the talks with the French shipyards. In the end he arranged, perhaps with Weiller's assistance, for a number of tankers to be built in the St Nazaire shipyards, with the French Government subsidizing 35 per cent of the total cost. He then made a deal with a number of American film companies which enabled Getty to buy French francs, which they were unable to export because of exchange controls during and after the Second World War, at 70 per cent of their face value in order to pay his 65 per cent share of the shipbuilding costs. The upshot was that, thanks to Forrester's charm and powers of persuasion, Getty paid only 45.5 per cent of the cost of the tankers, a remarkable coup.

The intriguing, and characteristic, aspect of this story is that not one cent of Forrester's $1 million finder's fee was paid by Getty. Forrester had to negotiate it with the French shipyard's management as a commission on the tanker-building contract. Forrester, says Buchwald, 'got healthy by way of a $1 million fee' which put him on easy street for the first time in his life. Buchwald thought Getty 'very sardonic, no one's fool, very fascinating and tight as Croesus'.

When it was all over Getty made a donation of $20,000 to the Mayor of St Nazaire. Paul Louis says, 'When he [Getty] was sure it was the right thing to do, he gave the money.' Getty himself received the Légion d'Honneur.

Getty contrived a similarly lucrative deal in Japan, where he arranged a $1 million discount on the regular tanker price by bartering crude oil and sugar to Mitsubishi, one of the great Japanese trading concerns, which also happened to be in shipbuilding.

The decade that spanned Getty's operation of his oil concession in the Middle East, from late in 1948 until the riproaring economic expansion at the start of President Eisenhower's second term in 1957, saw the multiplication

GETTY'S HOLDINGS IN 1957

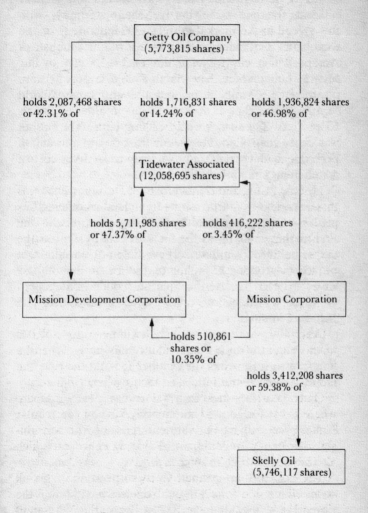

of the family fortune. The shares of Getty Oil Company, owned by Getty and the Sarah Getty Trust and worth $80 million at the time of the abortive sellout to Sunray, were now valued in the market at over $300 million. This was a far better performance than the Dow Jones industrial average, which was driven up almost 300 per cent by the postwar bull market. The Sarah Getty Trust was now worth over $200 million, or almost seventy times what it began with in 1934.

But Getty Oil shares worth $300 million were not all that Getty controlled one way or the other by the fall of 1957. He had created his own empire of interlocking shareholdings to control Tidewater Associated and Skelly Oil, Mission Corporation and Mission Development Corporation. This financial structure added another $350 million to the paper wealth that Getty controlled. The total market value of all the Getty related stockholdings was some $653.39 million. These holdings do not take into consideration the value of the Pierre Hotel, the Mexican beach resort, the Spartan mobile home operation, the art collection, or the value of oil reserves, if they were liquidated.

Getty must have had a pretty accurate picture of his paper wealth for some time, and it cannot have come as a complete surprise when, in October 1957, at the ripe old age of sixty-five, the hitherto obscure oilman burst into the public consciousness in a way that was to change his everyday life for good. That monthly bible of capitalism, Henry Luce's altar to private enterprise, *Fortune* magazine, decided to tell the world that J. Paul Getty, the 'California expatriate now living in Paris', was America's richest man with an estimated net worth of $700 million to $1 billion (i.e. one hundred thousand million). The Rockefellers, the Mellons, the DuPonts, the Whitneys, the Astors, the Harrimans, even Mrs Frederick Guest, his one-time landlady in New York, must have been surprised. He was the richest American, richer even than Howard Hughes, another eccentric character with a penchant for nubile beauties; richer than Joseph P. Kennedy, who was using his wealth to build a political dynasty through his sons; richer than the mysterious Daniel K. Ludwig, the shipowner, each of whom was estimated to be worth only $200–$400 million.

Getty did not know how to respond to the news. It was as if, overnight, he had been transformed into a cult figure. When he heard about the *Fortune* article he told his brother-in-law, Ware Lynch, 'I don't know how much money I have, I don't know how they would know. Have them see my attorney.' In fact, the article was probably based on the information he had himself given to *Fortune* in the first place. Each of 175 American multimillionaires thought to have more than $50 million had been sent a long questionnaire to answer and then fifty of them had been subjected to a long interview. It would hardly have been in character for Getty to turn down the opportunity of top billing. After all, he told Art Buchwald, soon after the article appeared: 'It was rather amusing in a way that the story should come out in *Fortune* because they've ignored me for twenty-one years. I kept reading about other people in the magazine with large businesses, and I always thought since mine was larger, they'd get around to me. Finally, when they did, they hit me right between the eyes. It looks like I'll have to change my name if I expect to get any peace.' He was probably not upset, or even mildly annoyed, to be at the top of the *Fortune* parade. He loved the publicity, and his reactions were full of contradictions. He told Buchwald that 'the news about being the richest man in America came to me as a surprise.' And, in the next breath, he said, 'My bankers kept telling me for the last ten years it was so, but I was hoping I wouldn't be found out.' Maybe he thought two lies were better than one.

Naturally the newshounds of Europe and America soon began to badger him, but in the best circles one does not crow about one's superior wealth and Getty tried to get his act together. 'Really, one has heard rumors,' he said loftily. 'I can't say it thrills me. When one is very rich it doesn't specially matter to be just that bit richer than someone else.' Still, he couldn't quite resist the boast: 'I've never really been able to count mine.'

Now that he was out in the open, he was the object of envious potshots from other wealthy men like Charles Wrightsman, a major benefactor of Washington's National Gallery, who tried to put Getty down with comments such as, 'Paul! Why he's not worth more than half a billion!' *Forbes* magazine decided to outdo *Fortune* by making him worth $1.6 billion, a figure that Getty never denied. Indeed, he was quoted in a London paper estimating his wealth at $1.7 billion. He subsequently repudiated this momentary lapse into American chauvinism and went back to being an English gentleman. 'It's vulgar to talk about money,' he said and was concerned that this boasting might 'get me blackballed from the London club I'm up for.'

Nevertheless, he did talk about his wealth to Ralph Hewins, his official biographer. He authorized for publication a list of assets, which no one else could have given to Hewins. These totalled $1,272,900,000. His tankers were valued at $200 million, while the art works at cost were only $4 million, and the Iranian oil interest was valued at only $1,100,610. Hewins had had to sign a letter promising that every page of the book would be checked. The trouble was that Getty wanted to be known as a billionaire, but he also wanted to wear a mask.

He tried every way to find a posture that might make

him seem comfortable with his wealth. Perhaps he should denigrate the very act of making it. He took the opportunity of doing so in the most widely read of all the many pieces about him, a cover story in *Time* magazine on 24 February 1958. He told them: 'The trouble is that everyone talks about how much money I make. I wonder what sort of accomplishment it is to make a lot of money.' *Time* also spoke to Getty's publisher and friend Mark Goulden, who said, 'I don't believe for a moment that he gets any enjoyment out of his money. He's just a miser – period.' Maybe not. All the same, Getty's countenance in the formal *Time* portrait was that of a well-satisfied man, a smile, crossed with a smirk, softer than the image that often emerged in other photographs which made him appear sardonic, devilish, or just plain angry. He looks wise like a cat, not wild like a lion.

For *Business Week* Getty put on a mournful countenance and explained modestly, 'I have never particularly wanted to make money, but as my business developed over the years, money came with it.' Without a doubt Getty was a candidate for Oscar Wilde's famous description of the man who 'knew the price of everything and the value of nothing,' an insult thrown at him countless times and on the slightest provocation by Mary Teissier, who became a verbal sparring partner. He had other worries too about the publicity he was receiving. 'I suppose I have to increase my tips,' he complained to Buchwald, 'from 14 cents to 35 cents.'

The only undeniable truth was that *Fortune* magazine had spotlighted a man who would be 'thenceforth a curiosity only a step or two removed from the world's tallest man or the world's shortest midget.'

15
Lord of the Manor

Why I like dogs better than people: dogs always like you for yourself.

J. Paul Getty

Once it became public knowledge that Getty was the richest man in America, it became apparent that he would have to give up living anonymously in a series of small suites in all-too-public hotels. Kidnappers could be lurking in the lobbies, beggars waiting on the doorstep, business spies rummaging through his wastepaper baskets, unscrupulous newshounds checking in to keep an eye on him and his companions. Now he needed a more permanent, more secure, less public residence in which to escape from his newfound fame.

By 1959 heavy taxes in Great Britain and the inability of some of the landed gentry to replenish the family treasury made many stately homes available to anyone with sufficient ready cash to help the owners make ends meet. Rich Americans in search of a proper ambience to disguise the novelty of their wealth were often the willing buyers. Opportunity knocked for Getty on 27 June 1959, when Paul Louis Weiller took him in his Rolls-Royce down to a dinner party at Sutton Place, the Duke of Sutherland's estate in Guildford, Surrey, some 23 miles from London. The Duke who had purchased Sutton Place in 1917 for £120,000 (more than $600,000 at the then rate of exchange), was losing a great deal of money on his farms which were vastly overstaffed, and he was in dire need of cash according to Albert Thurgood, the estate manager of Sutton Place. Getty, who was never one to miss the opportunity of a bargain even if it meant taking advantage of someone else's troubles, decided that the famous manor house could be had for a song. He offered the land-rich, cash-poor Duke only £50,000 ($140,000) for the house and about 60 acres of parkland. Weiller was embarrassed that his friend had deprived the Duke of his home 'for the price of two swimming pools'. Getty 'conservatively estimated' his purchase price 'at less than one twentieth its replacement value', and regarded it as an even better deal than the Pierre Hotel and most of the art work he was buying.

Acquiring a stately home in England had other advantages, he was to discover, apart from seclusion and privacy. It was going to be a lot cheaper to live at Sutton Place than in the Ritz Hotel, where he was chatting one night in October 1959 with his friend Art Buchwald. Their drinks – a glass of tomato juice and a rum and Coca-Cola – would cost only a dime at Sutton Place, he told Buchwald, instead of $1 at the Ritz. Dinner, he claimed, 'shouldn't cost more than $2.50 a head. But if you had to take six people to Maxim's it would cost you $10 a head.' Another advantage Getty recognized was the comparative cost of servants. 'You could get a dozen servants in England for what three cost in the United States,' he assured Buchwald. The result of their talk was a column in the *Herald Tribune* called 'Hotels Cost Money'. When it came to the point, Getty was no more trusting of his servants than he was of anyone else. He would 'lock everything up and keep the key.'

On the day that the deeds to Sutton Place passed to Getty a photograph was taken of the new owner and a group of his friends. They make a slightly bizarre, higgledy-piggledy collection, without any unity, and the mood is decidedly not upbeat for such an auspicious occasion. On the right, looking uncomfortable, even uneasy, his mouth twisted in a combination of grin and grimace, is Getty himself. On his left is the sixteen-year-old Jeannette Constable-Maxwell, whom he had met at Belvoir Castle, the Duke of Rutland's home in Berkshire, and whose coming-out party was to be the official opening ceremony of the J. Paul Getty reign at Sutton Place.

Three of his lady friends posed for the photograph – Penelope Kitson, a tall, British giraffe, dressed plainly, smiling uncertainly, and Mary Teissier, still beautiful and smartly dressed, looking not at Getty but at the third of the ladies, Marian Anderson, who looks proudly at Getty, the nomad turned into a landed aristocrat. Close to Mary Teissier is Gordon Getty, the fourth son, standing casually in sports coat and slacks with hands in pockets. In the background, smiling sheepishly, is Norris Bramlett, who kept the books of the Sarah Getty Trust and the Malibu ranch and museum and looked after the obligations of Getty's life back in California. Bramlett is dressed in a ready-made suit several sizes too big, with a fedora on his head hiding incipient baldness, and a tie that might have come from former President Harry Truman's Kansas City haberdashery. Finally, over Mary's shoulder, can be seen the rugged outline of Everette Skarda, the Getty Oil geologist, back from the Middle East. The picture reveals a ménage which is about to take over Sutton Place and which is very different, one feels, from that presided over by the impeccable Duke of Sutherland.

Sutton Place is described in *Annals of an Old Manor-House*, a book devoted exclusively to it, as a 'building of singular interest in the history of art, as well as of a rare and peculiar beauty.' From the outside, it is a graceful structure consisting of a main building with two wings made of reddish-brown brickwork and terra-cotta. Inside it is grander, with more than a dozen reception rooms, including the so-called Long Gallery measuring 165 feet and said to be the largest hall in Britain. Getty could seat nearly a hundred people in the dining room at the 70 feet of refectory table which he purchased from Hearst's Welsh castle. There were fourteen principal bedrooms, together with servants' quarters and pantries. In the garden were yew trees sculpted to resemble flowering tulips. Sutton Place is mentioned in the Domesday Book, and from 1066 onwards it passed through the hands of various nobles until King Henry VIII gave it to Sir Richard Weston, one of his favourite courtiers, in 1521. Weston built on the site the first non-fortified Tudor mansion in southern England. Several of Henry's wives had graced the sombre, oak-panelled great hall that was to become the entrance to Getty's home. Weston's son, Francis, was beheaded for allegedly bedding Anne Boleyn one night, but Sir Richard rose steadily in the King's favour. Perhaps he had been as astute with numbers as his successor Getty five centuries later for he was appointed Under-Treasurer of England, and he lived to entertain the future Queen Elizabeth. Interestingly, it was Alfred Charles William Harmsworth, later the newspaper baron Lord Northcliffe, who leased Sutton Place for many decades before passing it on to the Duke.

What became known far and wide as Getty's home had actually been purchased not by Getty himself but by a subsidiary of Getty Oil Company called Sutton Place Properties. The legal side was handled by Getty's British lawyer, Robina Lund, the daughter of the Law Society director, Sir Thomas Lund. Then in her mid-twenties, she was a personal friend of Getty and became not only one of his great admirers but also a director of the new subsidiary. Sutton Place was to be designated Getty Oil headquarters in Europe, an extremely important manoeuvre.

Establishing Sutton Place as an American subsidiary kept it from the grasp of the Inland Revenue. Income tax was as high as 98 per cent for the very rich in Great Britain, but the maximum was only 70 per cent in the USA. There was even a formal assignment of Getty as chief executive officer to be resident in the UK to protect him against claims by Her Majesty's Government that his annual income should be subject to confiscatory British taxes. Getty may have been resident in the UK, but for tax reasons he was domiciled in the USA.

The grand opening of Sutton Place was celebrated by the debutante party of Jeannette Constable-Maxwell. With her dark eyes and red mouth, she may have reminded Getty of the teenage girls he had courted in California in the 1920s, but she was definitely not a girlfriend. Sophisticated, intelligent, slightly fey, an elusive young woman who had lost her mother early in life, she was the daughter of Getty's friend Ian Constable-Maxwell, a Scottish Fraser and related to the Duke of Norfolk. The invitation read: 'Captain Ian Constable-Maxwell requests the pleasure of your company at Sutton Place on Thursday 30th June for Miss Jeannette Constable-Maxwell at the kind invitation of Mr Paul Getty. Dancing 1:30 o'clock. White tie.'

For the party, Sutton Place was lit up like a palace on Coronation night. Gold-painted standards bearing flaming torches lit up the outdoor pool area, and the swimming pool itself was lit by lights under the water. Another unusual touch was a prize Guernsey cow called Jessie, provided by the Milk Marketing Board, which moved among the guests, inside as well as out, offering non-alcoholic beverages. This novelty had been provided by one of Getty's new aides, Claus von Bulow, who was mainly responsible for making the party go. A Dane by extraction and a lawyer by training, he was a tall, ruggedly handsome man with a regal, but sinister, look and had added the 'von' to Bulow to give himself added panache in international society. He had been trained in Lord Hailsham's legal chambers and recommended to Getty by his tutor in how to get on in society, Paul Louis Weiller. Von Bulow had 'literally moved into Sutton Place for a while, and served in many capacities, not the least to amuse Getty at meals.' He found that 'Paul was a man of great, low-key humour' and felt only 'admiration, love and gratitude' towards him. Getty gave von Bulow '*carte blanche* to pad out his list of partygoers', and Weiller added to the list a few special titled people with whom Getty should curry favour. But not a single one of his sons was present.

Von Bulow certainly helped to create the party of the London season. The guests included an assortment of dukes and duchesses like the Gloucesters and the Rutlands, glamorous film stars like Douglas Fairbanks Jr and his exquisite daughter, Victoria, Greek shippers like Stavros Niarchos and Aristotle Onassis, and Sam Spiegel, the American film producer. Though Sir Winston and Lady Churchill declined the invitation, their daughter and son-in-law, Mr and Mrs Duncan Sandys, were there to represent them. Getty was supported by a full assortment of his close women friends, Penelope Kitson, Mary Teissier, Robina Lund wearing a dazzling tiara, even Ethel LeVane, his co-author and travelling mate, as well as Madelle Hegeler, a stunning model from Paris. A photograph of him that appeared in most journals shows a sombre old man with a long fleshy nose, a taut mouth and staring eyes, impassively leading around the dance floor a tall, young beauty with five strands of pearls on her rounded bosom. There was no sign of animation or enjoyment, but there was no need for the apparent sense of strain. Not since impresario Mike Todd, the former husband of Elizabeth Taylor, had transported the Battersea Fun Fair into a gala which he threw for the British party-going set had an American made such a social impact.

Jeannette Constable-Maxwell barely recognized half the guests because she had never met many of them before. (Nor had Getty for that matter.) 'There were troops and troops of strangers,' she says. One thousand two hundred guests had been invited, but more than twice that number came, some by gatecrashing, others by buying fake invitations on the black market at astronomical prices. According to von Bulow, William Hickey, the *Daily Express*'s gossip columnist, smuggled three assistants into the party as fake waiters. In fact, anyone could get in because at one point a line of six hundred cars was blocking the London road, and the security simply broke down.

When it finally got under way the party was, by all accounts, a roaring success. It took the greater part of two and a half hours for the horde of guests to shake Getty's hand and kiss Jeannette on the cheek. A fourteen-piece orchestra began playing at 1.30 A.M. At 6.15 A.M. it was still going strong. The next to last word on the party's success came from Getty's friend, Jack Forrester, who told him, 'You got a fantastic bargain, Paul. It would have cost your companies fifty million dollars at the very least to buy the same amount of advertising space.' The last word of all was spoken by Barbara Wallace, who had just become Getty's secretary and was to be privy to his inner and outer life until his death. 'It was not just a party for Jeannette,' she says. 'It was Paul's coming-out party, a way of introducing him to society.'

One might say that Getty had arrived, at least in relation to his 'image' which meant so much to him: a basically simple, but incredibly rich, Midwestern oilman, living in a stately English home, unmarried (for the fifth time), with a remarkable reputation with women, courted by dukes and duchesses on the one hand and, on the other, by hard-hearted businessmen such as the chairman of Royal Dutch/Shell and British Petroleum, who treated him with all the deference and respect which he felt was due to his impressive wealth and power.

There were too, of course, the crowds of individuals everywhere who saw him as the repository of their hopes, and who wrote to him expecting him to bail them out of some financial or personal problem. How were they to know that Getty did not believe in charity? As he himself put it, 'You should use the money for your own business and not for charity.' His charity, he insisted, consisted of the '10,000 people dependent on that business, as well as a government that receives taxes from it. If I make the business successful, I think I'm doing a great deal for the public good.' Or so he told Buchwald as they sat in the Ritz Hotel. Of course, he was not exactly being four square with Buchwald about all the women on the receiving end of his largesse when he swore, 'I never give money to individuals. It's very unrewarding and unscientific.'

If Getty was known for his wealth, power and glorious parties, he was soon to become better known for the biggest gaffe that he made – the installation in Sutton Place of a telephone with a coinbox similar to those in public places everywhere in the country. There was even a sign 'Public Telephone' on the wall of an easily accessible room on the ground floor. Special locking devices were placed on all the other Sutton Place telephones. Before this was done the quarterly telephone bill was surprisingly large and Getty could never make up his mind whom to blame. At first, he blamed the workmen restoring Sutton Place who, he suspected, were calling their relatives in Canada. Later, he claimed, 'people attending parties, charity functions, or open-house receptions at Sutton Place were casually picking up the telephones they found scattered around the house and placing calls here, there and everywhere.' A call from Sutton Place to London cost the mighty sum of 6d (18 cents), but, as Getty complained to the press, 'When you get some fellow talking for ten or fifteen minutes, well, it all adds up.' His old acquaintance William Randolph Hearst, he recalled, would send a weekend guest packing from San Simeon if he or she dared make a long-distance phone call without his knowledge. Getty even let the ignominious telephone box be photographed, as a demonstration, perhaps, that he was not just a rich American who was a light touch for his British guests.(His aides working in the basement of his home in California in the late 1930s had to use a payphone to make their personal calls too.)

When the barrage of publicity became too severe, he appeared to back down. The phone was being taken out, he said, as the workmen had finished their restoration. All the same, it remained in the downstairs cloakroom, where it was hardly ever used. The pay telephone was a story on which Getty kept ringing the changes. In 1965 he gave a new version to the *Saturday Evening Post*. 'There was this business of the payphone I had installed in my country house in England, to be used by my guests. When I'm staying with friends and have to make long-distance calls, I make a point of making them from a payphone in the nearest town or village. I had the payphone installed in my place because I knew that guests preferred it that way. It saved them the trouble of settling with me afterward, or of attempting to pay for their phone calls.' As it happens, no personal guest can ever seem to remember having paid for a phone call from Sutton Place, although Mary Teissier told Lady Diana

Cooper that 'Paul complained about her weekly calls to her mother in Paris.'

His admirers felt that the miserly act was just that – an act. Just as F.D.R. had his jaunty cigarette holder and Churchill his big cigar, so, according to von Bulow, Getty had his lugubrious face: 'His parsimony was to some extent an act, a funny feature like Mr Rockefeller distributing dimes.' The wise, sardonic Mark Goulden came to a 'startling conclusion' about his friend's 'studied, deliberate, unnecessary and calculated parsimony'. While recognizing that Getty was driven to increase his wealth, he felt 'in a way it was an abstract thing that didn't somehow belong to him; a kind of sacred, priceless orchid not to be touched and from which no petal must ever be plucked.' In short, Getty's wealth represented his duty, some deep responsibility to his father and mother.

He could never make up his mind what image he wanted to present to the public, though 'he wanted to project an image different from the one of miserliness and frugality.' Or so he told B. W. von Block, a ghost-writer whom he had hired to write his autobiography at what was, in 1958, an amazingly high price of $50,000. Bela von Block had a problem because Getty 'was generous as long as no one knew about it.' He also found it difficult to smile. Even when he told a joke he hardly ever changed his expression and his mouth would tighten. According to von Block, Getty's frozen face, the image of a tightwad, was caused by the fear that he might be photographed in an unguarded moment and the caption might then read: 'What secret deal has J. Paul Getty just completed to make him gloat with such unabashed glee?' The cartoons of the 'richest American' made him look a sad old grouch, a Scrooge, a bloodhound on the scent of even more money, but never enjoying himself.

The problem he faced was how to deal with having all

this money. People wanted you to give it to them. Or
they wanted you to tell them how to make it. On the
other hand, in the social milieu which Getty admired
money talk was considered rather vulgar, a typically
American barbarism. There were several ways of dealing
with the problem. One was to be tight with the servants.
The gatekeeper, he told Mary Teissier, was worth only
£2 10s ($5.60) a week, a figure he arrived at by multiplying
the number of times the Sutton Place gates had to be
opened or closed. A second rule was not to talk about
money in front of the nobility. 'He would talk about art,
about museums, so he wouldn't have to talk about
money,' says Claude Sere, a French furniture dealer who
met Getty through the Duke and Duchess of Bedford.
'His attitude towards money was a disease.' The Duke of
Bedford, who was as close a friend as Getty had among
the top nobility, recalls advising him on how to treat
various classifications of guests at Sutton Place. 'There
were set meals for the A, B, C and D groups,' he says.

Charles Lee, his admiring chauffeur, found Getty far
more approachable and friendly than the Duke of Sutherland, his former master, who treated Lee simply as a
servant. Lee 'never had a cross word from Getty, who
would walk in unannounced to my living room, sit down
and have a cup of tea.' Still, Lee found him 'a very
strange man' who would ask him to stop for a moment at
the British Museum and Lee would wait outside for six
hours while 'Getty was still there, hunched over his book
with his glass on, reading in a cubicle.' Sometimes Lee
would take him to Annabel's night club, the in-spot for
posh upper-class Britons, where Getty would stay until 4
A.M. dancing with the Duchess of Argyll.

Getty would drive miles to see his favourite film, *Queen
Christina*, starring Greta Garbo. It must have amused
him to see Garbo masquerading as a man and perhaps it

struck a chord in him when she rolls her beautiful eyes to the sky about the 'duty, duty' which requires her to stay on the throne of Sweden.

Despite his reputed search for privacy, more words were printed about 'the richest American' – or by him – than about any other very rich man alive at the time. Ralph Hewins, in his obsequious biography *The Richest American*, made excuses for Getty's foibles. He promised Getty in writing not to criticize any Arab nation and swore he himself was not a Zionist.

Then there was the autobiography *My Life and Fortunes*, ghost-written by von Block, who got in to see Getty at the Ritz by sending him a five-page telegram. *Newsweek* said of the autobiography that it 'turned a flat minimum of soul-searching into a bland pudding of self-justification and pious maxims.' But it was in a series of articles in *Playboy* magazine that the Getty mythology was really spread. They were put together and ultimately published in two paper-back volumes called *How to Be Rich and How to Be a Successful Executive*. In them Getty signposted the path to success and riches: be a nonconformist and avoid like the plague an 'accumulative mentality'. He seemed to be criticizing his own strongest attributes.

Executives' morbid preoccupation with their health 'is a byproduct of the status-seeking mania', he wrote. His own bathroom was filled with the world's largest inventory of pill boxes, bottles, ointments and prescriptions of all kinds. He would not let a person with a common head cold near him for fear of catching some disease that might stop a business deal in its tracks. He let the BBC film him at seventy lifting his barbells first thing in the morning and reading Henty's boys' stories of adolescent heroism to himself at night. He was also filmed at the Duke of Bedford's estate, Woburn Abbey, chortling about the Raphael painting he had purchased for only $200 in 1938.

But in his introduction to *The Joys of Collecting*, a work about his art collections, he wrote that 'the true worth of a collection cannot – and should not – be measured solely in terms of its monetary value.'

Perhaps the most absurd article to appear under his name appeared with the title 'The World Is Mean to Millionaires' in the *Saturday Evening Post*, 22 May 1965. The reader is encouraged to pity 'the richest man in America' suffering in his castle over there in Great Britain. 'Rich people once lived in a world apart; today almost the only difference between the multimillionaire and the reasonably well-to-do man earning $15,000 to $25,000 a year is that the millionaire works harder, relaxes less, is burdened with greater responsibilies and is exposed to the constant glare of publicity.'

Unlike most corporate chieftains, who try to guard their private lives and present either a simple, bland or a distinctive, colourful image to the world, Getty went out of his way to appear as a mass of contradictions. At the same time grandiose and humble, public and private, a loner and a socialite – it was as if he were content to be a kind of curiosity and therefore in some way larger than life. But beneath the surface? His friend the Duke of Bedford did not think 'there was a surface to get under. He was pleased by the publicity.'

Certainly he had his share of that. Not only was there the payphone to intrigue the newspapers. There were also awesome stories of the security at Sutton Place, which included patrols of dangerous-looking Alsatian dogs. These had been provided by Leon Turrou, a former FBI agent, now based in Paris and author of a book on the Lindbergh kidnapping. Visitors were unnerved by a sign which read: 'Danger. Guard Dog – Keep away. This dog is trained to treat all strangers as enemies. Do not touch. First-aid kit in cloakroom tent.' The dogs, in fact, would attack just about everyone except Getty himself. His chauffeur, the cheerful Charles Lee, had to move from the servants' quarters to a cottage when his wife was mauled by Shaun, the master's favourite dog. Shaun had the run of the mansion, and used to terrify guests by poking his fierce snout over the edge of the table. John Houchin, the president of Phillips Petroleum, thought the look on the dog's face meant 'Give me something or I'll give you something.' Turrou also advised his master to bar the famous stained-glass windows of Sutton Place to protect himself and his works of art. It kept his insurance premiums down, Getty explained.

Many visitors found Sutton Place barren of fun, cold, institutionalized, a museum where the curator and his small band of attendants lived grandly but quietly. The massive oak doors barred with ancient bolts reminded Helen Lawrenson, a writer for *Esquire*, of an old Elizabethan dungeon. Buchwald told his wife about 'this lonely old man living in this huge place with one or two other people.' An upstairs maid remembers the only sounds of life as the hoovering of rugs and the chatter of the secretaries' typewriters. It must have been very much like the austere silence at his former home on South Kingsley Drive, when he lived there with his mother who grew more deaf every day.

Even his friend the Duke of Bedford never had any fun at Sutton Place. 'It was an ordeal more than anything. He didn't offer much in the way of conversation. He didn't have the key to real happiness – an ability to share.' On the other hand, he often invited himself to Woburn Abbey and would sleep 'in the staff room as long as the bedroom door could be locked and he had a bottle of mineral water inside. Paul was the easiest weekend guest you can have. The children would teach him the new dances, and he would sit on the carpet and play Monopoly.' Nicole Bedford, the Duchess, believes that old-world courtesy was Getty's special quality. He once wrote to her 'the most beautiful note I have ever received', which simply said, 'Five miles before I arrive at your house, I start feeling happy.' One morning, however, the Duchess found him 'jumping up and down like a yoyo. I asked him what the matter was. Penelope could not sleep all night, as the door kept on opening, and she felt cold hands on her face, so I told him about our friendly ghost. Paul asked to switch bedrooms with Penelope, and spent the whole night waiting up for the ghost, which never came.'

Beneath the apparent quiet of Sutton Place, however, a strange kind of drama – or comedy, according to how one sees it – was unfolding in which every member of the inner circle took part. It was a version of Schnitzler's *La Ronde*, and its theme was that 'everyone was vying to be mistress of Sutton Place,' as Jeannette Constable-Maxwell, a friendly observer, put it. 'Paul Getty was buffeted by the people around him.' Jeannette was fond of the older man, who was roughly the same age as her father, but she knew that one could not live the way he did and expect to be truly loved. She saw him as 'someone who needed help, unique, a true friend, a pal. He was not sure of his values. He lacked the knowledge of real values.'

His adoring secretary Barbara Wallace thought 'his weakness was people. He was incredibly vulnerable and soft. Yet the magnetism got you. You just felt you had to do things for him.' This combination of power and vulnerability, of unlimited money and the need of a child looking for a mother, appealed to a whole series of women, both those who were his lovers and those who were just friends.

Penelope Kitson, the interior designer of Sutton Place,

was ensconced in a small cottage called the 'Cricket Pavilion', after its use in bygone days. By her own admission she led her own life. According to von Bulow, although she 'didn't want to be the sixth Mrs Getty, he trusted her and was right to do so.' She also recognized that he was 'not a very good judge of character, was easily flattered, and was frightened by being alone.' Whether in a Paris night club, or riding a camel on a Middle East trip she made with Getty, Penelope was elegant and straightforward and had her head firmly planted on her shoulders. She was 'a real friend, and he knew it,' says von Bulow. Apparently, she knew the right balance between being understanding and helpful without being too bossy.

Mary Teissier competed with Penelope for Getty's favour. She kept her personal belongings in a bedroom at Sutton Place which had a communicating door with Getty's. She hung a portrait by Pietro Annigoni of herself as a still ravishing beauty in the study, where the other women were bound to be irritated. Mary's connection with the Russian royal family was through the Grand Prince Alexis, who committed the social gaffe of marrying his mother's lady-in-waiting, thereby losing his royal status. He was packed off to America, and Mary's mother was one of his several children. 'Mary became a prisoner, and she took it out on Paul in the most silly way,' says her cousin, Alexander Romanov. The scenes between them became more and more absurd. She would interrupt him and blow smoke in his face. He would simply accept the abuse and treat her with his own brand of aloofness.

Their affair was a passionate one, by all reports. Mary would use ten different perfumes on her body, daring Getty to identify each one, according to Mary's close friend Ariane Pathe, writing in an Italian magazine. Getty only made love on his king-sized bed, according to Pathe, because 'he was turned on by the thought that that was where Henry VIII possessed Anne Boleyn for the first time.' Mary told Pathe that Getty said to her, 'Nobody in the world will ever give you the physical pleasure I can give you.'

Getty's reputation as a lover presents something of a mystery. He was never seen to be physically affectionate with any woman. There were no embraces in public, no arms around the shoulders, no hand-holding, no signs of romance. He was reported to eat spaghetti in bed with Mary before making love. While Mary bragged about his lovemaking, she also said, 'I loved Lucien for his body and Paul for his mind.' The Duchess of Argyll, who was not one of his lovers, calls him 'a guru, a Buddha', and von Bulow felt that he 'was not a ferocious indiscriminate womanizer'. However, he liked the company of several women at any one time. If a new female conquest sighed, 'Oh, Paul, this is hopeless,' his casual answer would be, 'Well, I should miss you, dear.'

Von Block, the ghostwriter, was told by several women that Getty would make certain that they had an orgasm, and both von Block and Turrou were awed by the size of his penis as well as his bank account. Turrou believed Getty was 'as well hung as any man I ever saw, maybe 8½ inches long in the swimming pool,' and according to von Block he was 'built as very few men are. He was noted for his sexual prowess.'

Getty himself felt that success in business and success with women were inextricably connected. He told von Block that 'business success generated a sexual drive, and that sexual drive pushed business.' Or, as Norris Bramlett puts it, 'Mr. Getty did not separate his business and personal life.'

Nevertheless, as he got older he clearly worried about his virility. For many years he took injections of H-3, a

Above left: Sarah Catherine Risher, J Paul Getty's mother, in 1899

Above right: George Franklin Getty, the original oil wildcatter in Oklahoma and California

Jean Paul at the age of twelve with his dog Jip on the porch of the Rightway Hotel in Bartlesville, 1904

An unsmiling Jean Paul (in the centre of the third row) poses with his primary school class in Minneapolis in 1906

Getty at the age of sixteen with his mother (right) and father and his cousin June Hamilton in front of the family residence on South Kingsley Drive in 1908

Above left: Jeanette Demont married Getty in 1923 when she was eighteen

Above right: Adolphine Helmle-Fini married Getty in 1928

Left: Ann Rork, who married Getty in 1932

Teddy Lynch
became Getty's
fifth wife in 1939

Getty in 1936 at the
age of forty-four.
Despite the
Depression, he was
one of the
wealthiest men in
America

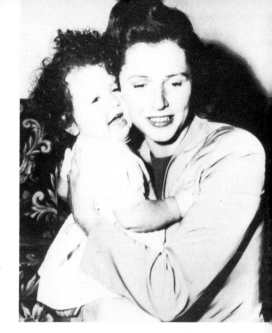

Joan Barry holding her daughter Carol Ann during her paternity suit against Charles Chaplin

With Mary Teissier, one of Getty's close friends in later life, a descendant of the last Czar of Russia

On his first visit to Saudi Arabia and his oil concession in the Neutral Zone, Getty dines with Saudi Dignitaries (1956)

Getty in 1955 dining with Penelope Kitson, another close friend from the mid-fifties until his death, and Prince Massimo

A rare smile at Jeannette Constable-Maxwell's coming-out party, which was held at Sutton Place in June 1960. Jeannette's father is on the left

Getty in conversation with his British lawyer and close friend, Robina Lund

An unusual pose for Getty with Penelope Kitson and Lady McIndoe (1965)

The day Getty bought Sutton Place in 1959. With him are, from left to right, Jeannette Constable-Maxwell, Penelope Kitson, Norris Bramlett (in the trilby), Gordon Getty, Mary Teissier, Everette Skarda and Marian Anderson

With Robina Lund at a party

Getty with his eldest son, George Franklin II, three weeks before George's death in 1973

Ronald, Getty's second son, aboard an ocean liner. Like his father, Ronald is afraid of flying

Gordon, Getty's fourth son, listens to Penelope Kitson

Getty's third son, Paul Jr. with Talitha Pol, who became his second wife

Teddy, Getty's fifth wife, and their son Timmy, who died at the age of twelve

Getty's grandson, J Paul III, surrounded by Italian police and detectives after being released by his kidnappers in December 1975

Getty with Rosabella Burch, a friend in his last years

The J Paul Getty Museum in Malibu, California

Walking in the grounds of Sutton Place with Marianne von Alvensleben and Norris Bramlett. A bodyguard armed with a shotgun stands behind a tree

Communicating with Nero, named after the Roman emperor

Claire Getty talking to her grandfather at Sutton Place in 1975

Talitha with Tara in 1971

Prominent on the social
scene, Gordon Getty and
his wife Ann at the Orchid
Ball in Los Angeles in 1984

Paul III with his wife Martine and his son Paul Balthazar. On the
right is Martine's daughter Anna

Gail Getty, Paul Jr's first wife, with their children, Mark,
Ariadne and Aileen

relatively harmless serum that contained procaine which, like novocaine, dulls pain. Getty believed it gave him renewed sexual energy, and Clive Mackenzie, his local doctor, who became a regular visitor at Sutton Place, considered H-3 'fairly innocuous and good for warding off old age, but not a hormone'. When Getty's hands began to shake, Dr Mackenzie did not think he had Parkinson's Disease, but let him take a medicine called Sinemet, recommended by Paul Louis Weiller, which reduced the outward signs of the tremor. 'Paul heard the side effect was to increase virility, so he kept a lifetime's supply of pills in the cupboard, and took one a day purely from the sexual point of view,' says Mackenzie. The doctor attended to Getty's precautions in respect of his sexual activity. Before Getty made love to some women, Dr Mackenzie 'had to examine them to make sure they were healthy and not passing along any disease. Also, I had to make sure they had some protection against pregnancy.' One lady told Mackenzie she didn't have any contraceptives, and added, 'It would be marvellous if I had a child by him.' When this was repeated to Getty his verdict was final. 'She went the next day,' says Mackenzie.

Evidently Getty felt he could keep his ladies in line by the effective use of money. During the decade of the 1960s he changed his last will and testament no less than eleven times, shifting around the amount of money he was paying them and planned to leave to them after he had gone.

The list of beneficiaries had changed dramatically since before he bought Sutton Place. Back in September 1958 the ladies designated to receive money were Ethel Le-Vane, of Hyde Park, London, $500 a month; Margarete Feuersaenger, of Berlin-Charlottenburg, $125 a month; and Hildegard Kuhn, of Berlin-Wittenau, $75 a month.

By 1963 the rivalry between Penelope and Mary had

been decided in financial terms for all time. Penelope, whom Mary called 'The Kitson', had been moved to the top of the list and was due to receive $750, a rise of 50 per cent from her previous $500 a month. Similarly Mary was upgraded from $400 a month to $500. Robina Lund, who was on the payroll as a lawyer and public relations representative, would be given 1000 shares of Getty Oil Stock – only half the number of shares he intended Mary to have.

By September 1965 Robina Lund seems to have risen spectacularly in the hierarchy and has pushed Mary out of second place. Robina has described Mary's attempted interruption of her games with Getty in the outdoor swiming pool. Mary screams, 'Robina, Robina . . . you are too rough and he is an old man.' Getty hears this and mutters to Robina, 'She makes me feel even older than I am – she'll only be happy when I'm in a wheelchair wrapped round with rugs.' Is that why Robina is now to receive 5000 shares of Getty Oil on Getty's death, the same as Penelope's legacy and more than double Mary's? What's more, she is now awarded $750 a month, the same as Mary, but still not as much as the $1167 left to Penelope. In November 1967 Robina must still be in favour, because she has been raised again to 6250 shares, still less than Penelope's 7500, but a generous bequest all the same.*

In addition to a seat on the board of Sutton Place Properties, Robina, according to her own record of events, had now become Getty's travelling companion in Europe. She claims that he was her 'chaperon', and that she was his 'mixture of daughter, sister, mother – and father confessor'. The 'chaperon' and 'father confessor'

---

* There were twenty-one codicils to his will. Details can be found in Appendix III.

are pictured in her book, *The Getty I Knew*, frolicking together.

Money, manor house, acceptance by high society in Great Britain – all this was not enough for Getty. He had to achieve more and acquire more. His thoughts turned to Italy, to the ancient civilization built by his heroes, Hadrian and Caesar Augustus. There, perhaps, he could act out his fantasy that he was the reincarnation of the Roman Emperor Hadrian. 'I would very much like to think that I was a reincarnation of his spirit and I would like to emulate him as closely as I can,' he told Robina. Besides, he would like to own several beautiful homes, like his friend Paul Louis Weiller. And so, for $566,000, he bought the ruins of a sixteenth-century Italian villa on the Mediterranean south of Rome, owned by Prince Ladislao Odescalchi, and spent $2 million restoring the stables that might have been designed by Raphael and the ancient frescoes found on the site.

Palo, as the estate was called, made little sense for a man who did not wish to live in a villa open to the sea. 'He was scared the pirates would come for him,' says Barbara Wallace. He hardly ever stayed there, and never spent a single night on Gaiola, a small islet in the Bay of Naples which he bought from Baron Langheim, a wealthy Austrian nobleman. It was here, according to legend, that one of the more sadistic Roman emperors used to throw his victims into the sea.

Getty spent barely an hour looking at his acquisition, but somehow, he felt, these purchases of Italian property might transform him into an Italian. He instructed Giorgio Schanzer, the tall, affable Italian lawyer who ran the Getty Oil Corporation in Italy, to find out how he could become an Italian citizen. As for Palo, his ownership led to his acceptance in the most esteemed social circles, including Count Campello and Marques Theodoli. These two gentlemen invited him to join the world's most exclusive club, Rome's Circolo della Caccia, the hunting club whose members belong to the families which produced popes, princes and rulers of separate Italian states as far back as the ninth century.

Giorgio Schanzer truly admired his friend Getty and felt protective towards him. The grandson of a former Italian Foreign Minister, and married into one of the noble Roman families, he knew that Getty would have more trouble becoming a member of the Caccia than he would becoming an Italian citizen. The usual qualification was the noble birth of all four grandparents. Both Schanzer and Frederico Zeri advised him not to allow himself to be nominated, thereby placing himself at the mercy of these penniless, snobbish and, above all, envious Italians with only a tattered title to their name.

When he was blackballed, as predicted, Schanzer believes 'he never recovered. It hurt him deeply.' Getty seems to have taken the insult philosophically, at least outwardly. 'Club members were very insistent that I join,' he said, 'and it seemed churlish to refuse. That the affairs of a private club should be made public is unheard of. It just wouldn't happen in England or America.'

If he could not take revenge on the Italian nobility, he could at least act out his frustration by drafting an Italian will, which is a characteristically mischievous document and reveals the delight he took in playing his women friends off against each other. Penelope, Mary, Robina and, later, Lady Ursula d'Abo were to be allowed to use Palo, if necessary at the same time, 'inasmuch as the property is large enough to permit simultaneous occupancy by several families.' All decisions as to their use of Palo were to be made by the majority vote of three of his sons, who must have wanted nothing to do with this scheme. The two rivals were to share responsibility –

Penelope Kitson to have power over Palo's interior decoration, and Mary to be responsible for the exterior of the building. Schanzer should perhaps have the last word on this bequest. He says, 'It was his wicked way to ensure that Palo could never be used by anyone after he was gone.'

16
The Emperor

I feel no qualms or reticence about likening the Getty Oil
Company to an Empire – and myself to a Caesar.

J. Paul Getty

As an American oilman, Getty was in a class by himself.
He controlled over 80 per cent of Getty Oil Company,
and Getty Oil in turn influenced the activities of Tidewater, Skelly, Mission Corporation and Mission Development. He earned from a variety of sources over $1 million
annually, an imperial sum in the 1960s.

For a quarter of a century, from 1951 until his death in
1976, he never visited the headquarters of his oil empire
in Los Angeles. His ministers had to come to him from
6000 miles away with their budget projections, requests
for capital expenditure, terms for a merger, or even with
their recommendations for the appointment of higher
executives. He never attended a single meeting of the
board of directors of Getty Oil or an annual meeting of
the shareholders. He was far removed from the usual
endless committee meetings of a corporate headquarters.
Yet no one could make a major decision without consulting the founder, and he was in a time zone eight hours
away from Los Angeles. 'He didn't have to compromise,
because he was so wealthy and so totally in control,'
says John Houchin, who was then president of Phillips
Petroleum.

Yet in his unique position as Getty Oil Emperor, he

showed insecurity in his choice of close aides, a lack of wisdom and compassion in the training of his sons for careers in the oil business, and a lack of vision when it came to understanding the influence of politics on the future course of the oil business back in the USA.

His inner cabinet consisted of a group of aides who certainly seem an odd assortment. Norris Bramlett, who had become his right-hand man, was a short, dry Texan, and had survived the Depression by working in a Civilian Conservation Corps camp. His first job with the Getty organization had been in Carlsbad, New Mexico, but he had moved to Los Angeles in 1937 to become an office boy at $75 a week. By 1956 he was in charge of the Neutral Zone accounts and, in the words of Everette Skarda, he was 'the most competent accountant you can hire for $750 a month'. He was a quiet type, a faithful, discreet assistant, very deferential to the Emperor he wished to serve. Early in his career he had had to turn in the stub of an old pencil before he could get a new one, and this kind of treatment conditioned him for the rest of his life. He would spend months in the hot desert collecting facts, or carry messages between the Emperor's palace at Sutton Place and the hired hands in Los Angeles or New York. Bramlett said he 'got along famously with Mr Getty. He could get angry and was a very complex man. He was not exactly lovey-dovey.'

Then there was Colonel Leon Turrou, the former FBI agent, who was Getty's adviser on security. He was also his general fixer of embarrassing situations in Paris, where he lived. 'A strange fellow, who played on Paul's fear of physical danger,' is how von Bulow described him. Turrou's business cards identified him as 'special Consultant to J. Paul Getty' and he received a small retainer for his advice on security. But, in curious fashion, Getty also let Turrou handle his negotiations for the purchase of steel pipelines from Mannesman and other European manufacturers. For years, so he claims, Turrou negotiated two special discounts from Mannesman. One discount he remitted to Getty Oil, the second he kept for himself. It was irregular, and probably an item the Securities and Exchange Commission would have loved to have known about. Turrou was bitter that von Bulow and Mannesman took his special commission away in the early 1960s, insisting that his negotiations had saved Getty hundreds of thousands of dollars. Each December Getty would give Turrou his Christmas present, $3000 in cash, the amount he could legally give without paying any gift tax.

Getty's chief executive assistant was the glamorous, socially acceptable von Bulow. No doubt the old man found his charm and wide range of social contacts immensely pleasing. He often sent him to the Middle East and elsewhere to represent him. Von Bulow was a member of the Neutral Zone committee which met three times a year to discuss overall relations with the Saudis and the Kuwaitis. 'Paul's refusal to smooth the way for his business by doing business in "Rome as the Romans do" . . . was hard on me in my negotiations in Japan or Saudi Arabia, and indeed in Washington,' says von Bulow. 'In all of these matters I accepted Paul's wisdom, and was privileged to learn from him and to work so closely with him.' Bramlett says von Bulow 'was highly respected in the company', but Skarda, the geologist, thought him 'a nonentity, a nut and bolt chaser'.

Although von Bulow was close to Getty, he did not exercise as much power as C. Lansing Hays Jr, the tall, caustic lawyer who took over Getty's affairs when David Hecht died tragically of heart trouble in 1959. Hays, who grew increasingly scurrilous and arrogant as the years wore on, usually chaired the monthly directors' meetings and ran Getty Oil's legal department from his office on

Wall Street. Von Bulow describes Hays as 'obsequious towards Paul, and towards the Crown Prince George. It was not a role that came naturally to his choleric and intolerant temperament. Lansing compensated for this by being quite exceptionally brutal and insensitive with everyone else.'

Not only were the executives in Los Angeles intimidated by the Emperor and his courtiers, they were also not very well paid by oil industry standards. Getty believed in a day's wages for a day's work and no mollycoddling with benefits such as generous pension plans or even stock option programmes. Getty Oil Company, throughout the 1960s, lagged far behind most American corporations when it came to employee relations. Getty was still stuck in the Victorian era. 'No wonder the corporate culture at Getty attracted plodders,' says a lawyer familiar with the Getty empire. 'They couldn't do anything without his approval. He had the power to make every major decision until the day he died.' 'I do it all myself,' Getty said, and that applied to giving his personal approval to items ranging from the investment of more than a million dollars to the size of the cans used for food sent to oil workers in the Neutral Zone. He calculated that flushing the toilets there with desalinized water cost 6½ cents every time someone pulled the chain, so he ordained that salt water should be used for the purpose.

According to Jack Sunderland, who became president of Aminoil in the 1960s, 'Getty didn't trust anyone. He had lousy people and he liked to set one person against another. He thought that was the way to get the best out of people. There was no general manager in the Neutral Zone – only four managers for each function. All four reported to him by mail, not by telephone or cable because it cost too much.' Getty instructed each manager in the Neutral Zone to write him weekly. It was an exaggerated example of his belief in the axiom 'a man is only as wise as the information he possesses'.

This determination to make the Neutral Zone his personal fiefdom was particularly irritating to the Saudis, as Getty only travelled there on two occasions. The American consul general in Dharan, Saudi Arabia, reported to the State Department that Saudi officials were unhappy at these 'attempts to operate with absentee management, and without delegation of important decisions to officers in the area.'

It may have been this lack of political insight which led to the faulty timing of his vast $600 million expansion programmes, involving the refinery, tankers and gas stations. So intent was he on becoming the eighth sister in the petroleum industry that he quite lost sight of political reactions to the brutal competition for gasoline markets. Independent oil producers in Texas, Oklahoma, Louisiana and California did not take kindly to the flood of cheap Middle Eastern crude. Their sense of outrage made itself known in Washington, and in July 1957 the Republican administration of President Dwight D. Eisenhower decided that voluntary import quotas on cheap foreign oil were needed to protect the domestic oil industry.

This was not good news for Getty, who had spent $200 million building the Delaware refinery. The import quota for Tidewater was so modest that it turned the new refinery into a $200 million white elephant. Indeed, the entire $600 million plan depended on Getty being able to use his tankers to transport the cheap Middle Eastern oil to the USA where it could be refined and sold as gasoline to American motorists. Tidewater officials complained vociferously, but at least it must be said that there is no evidence that Getty himself used any political influence he might have had to get the Tidewater quota increased, although he had been one of the wealthy contributors – along with Bob Hope, Bing Crosby, James Stewart and Clark Gable – to Eisenhower's 1956 re-election campaign.

To be fair, Getty's import quota, far less than the Delaware refinery could process, was based on the output of the much smaller and outmoded refinery in Bayonne, New Jersey, at an earlier period, when Tidewater's other eastern refining operations were shut down and the new refinery was under construction. Tidewater was unsuccessful in convincing the government 'to permit it to process at least as large a percentage of the foreign crude as its competitors on the East Coast,' according to the Tidewater 1957 annual report.

Someone – perhaps it was Getty himself – decided the solution lay simply in ruthessly disobeying the voluntary quota system. After all, it *was* voluntary, so no law or regulation was actually being broken. The risk was small. But at a Cabinet meeting on 3 March 1958, President Eisenhower charged that the quota programme was crumbling largely because of the 'unwillingness of Tidewater Oil to cooperate in the voluntary plan', according to the minutes of the meeting. He told Commerce Secretary Sinclair Weeks and Treasury Secretary Robert Anderson to come up with a solution. He was not yet ready to declare war on Getty and the other quota breakers, but before March was out he 'reaffirmed his dislike of a situation where a recalcitrant company could deviate without punishment from a generally agreed program.' The disobedient oil magnates were saved briefly by Secretary of State John Foster Dulles, a former Wall Street lawyer whose firm defended giant oil companies, who persuaded the Cabinet to maintain the voluntary quota system for the time being. Nevertheless, on 10 March 1959 President Eisenhower, by means of a presidential proclamation, established a mandatory quota programme that put an end to Getty's grandiose scheme of using his Middle Eastern production in the US markets. Eisenhower's decision had disastrous economic repercussions, because it meant that the USA had to rely on expensive domestic oil during the sixties, at a time when national security was not under threat by a temporarily weak oil industry, as was claimed. National security was to be threatened later by dependence on the Arab oil producers when, in 1973, the Yom Kippur War triggered an oil embargo and a worldwide energy shortage that had dire effects on the world economy.

The mandatory quotas had an equally horrendous effect on Getty's tanker fleet. Built at a cost of $207 million, it had lost its *raison d'être*. Not only was there a surplus of tankers in the world market, but fierce gasoline price wars broke out on the West Coast, costing Tidewater a substantial share of its market there. From 1960 to 1963 alone Tidewater's share of the gasoline market fell from 6.9 per cent to 5.7 per cent. Getty had wanted to build a gas station on every corner, but he found that Mobil, Texaco and Shell 'fight desperately to maintain their position. You advertise, they advertise more; you build filling stations, they build filling stations.' He told *Fortune* magazine: 'I think business has to be guided by military history. You've got to plan campaigns and strategies . . . You've got to plan for all the things that can go wrong.' He had learned the hard way that his plans were too optimistic, so he decided not to 'get too widely spread out. Don't try to advance too rapidly. The thing in business is to distinguish between the possible and the impossible and stick to the possible.'

He had gambled over $600 million, loading Tidewater's balance sheet with more debt than ever before in history, a violation of his father's sacred law against taking risks.

The obvious thing to do now was to get rid of the uneconomic parts of the business, where competition was stiffest, and pare down that debt.

In the meantime, however, he had neglected one of the chief requirements of historically successful emperors – the need to provide a dynasty or at least a proper succession to the throne. He spoke of his love for his sons but, as with his parents, it was at best an ambivalent love and at worst malevolent. He confessed he had always 'found it difficult to be openly demonstrative with my boys.' Still, he 'hoped and desired that all my sons would enter the family business, eventually taking over from me just as I had taken over from my father, who founded it.' Indeed, for a time in the early 1960s he employed all four of his living sons in the oil empire. But when it came to the point, his dynasty was a group of sons squabbling among themselves and often with their father, an alienated group that could barely call itself a family.

'Paul browbeat them to death,' says Dave Staples, former president of Tidewater, who had all four working under him. 'They grew up without any strong direction. They were moving up too fast, riding on their father's coat tails.' This was no way to build a dynasty. Other empire builders bring their sons in close contact to learn the tricks of the trade in preparation for passing on the crown. But not Getty. Deep down, he did not seem committed to their success. He even relished their inadequacies, seeing them as only tiny chips off the old block and not the real thing. Norris Bramlett put it in a nutshell. Getty, he said, was 'illiterate when it came to being a father'.

Paul Jr remembers that when he was in high school his father returned a letter Paul had written to him with all the grammar and spelling errors corrected but with no further comment. 'From that moment on it was impossible for me to write him another letter,' Paul says now. 'It hurt me so much I never got over it. For ever after I wanted to be judged as a human being, not as an employee. I could never get that from him.'

The old man spent hardly any time with his sons, treating them all coolly and often sadistically. It is a wonder that any of them stayed and tried to cope with him but, unlike the wives, they could not sue for divorce. All four wanted to prove something to him, but nothing really worked. They wanted what they saw as their natural 'birthright', a share of the Getty fortune. He explained later that their lack of success in business was because 'children from a broken marriage will be brought up and educated to be totally unlike their fathers'. Nothing could have been further from the truth. If he had not abandoned them, he could have played a positive role in choosing their education and in offering advice and support. Because they were the sons of different wives and were competing for attention from their father, no lasting brotherly affection ever developed.

The oldest, George II, who achieved the greatest success in the oil empire, became president of Tidewater Oil at thirty-six and heir apparent to the Emperor's throne. The son of Jeanette, Getty's first wife, he was an intelligent, plodding executive, who suffered intensely from a deep insecurity fostered by his intimidating father whom, as a boy, he barely knew. He was emotionally closer to William Jones, his stockbroker stepfather, to whom he dedicated a building at the Webb School in Claremont. Still, George felt his father was 'the smartest businessman I know. Coming to see him is like a visit to Mount Olympus.' No matter how hard he tried to please, no matter how many votes he had as executor of the will, George felt 'I could never win'. When Tidewater's exploration for oil in both Pakistan and the Spanish

Sahara turned out to be expensive failures, Getty simply 'dissolved the foreign exploration division of Tidewater without asking George's opinion', according to Everette Skarda. Actions of this kind reinforced George's frustration at being president of Tidewater but not being in control.

Physically, George was the shortest of Getty's sons. He could look pudgy in his Bermuda shorts and he did not have his father's flair with women (or his half-brother's, Paul Jr). Perhaps his insecurity made him pompous, and he was considered boring by his father and by many in the Getty camp. Yet he had many qualities which his father lacked. During his duty on General Douglas MacArthur's staff prosecuting Japanese war criminals, George had recovered a wristwatch from a Japanese prison guard and returned it to the six-year-old son of an American soldier. He served as a fundraiser for the Los Angeles Philharmonic, the kind of civic responsibility that Getty in his whole life never assumed.

The love–hate relationship between father and son had its roots in the bitter competition that had existed between Getty and his own father, George's namesake. On the one hand, George was given two votes as his father's executor to one for each other brother, and was appointed director of the Getty Museum, a quite amazing concession. On the other hand, he resented having to consult his father each day by transatlantic phone before he could clear the company business. He wanted to be left alone to manage Tidewater and one day wrote to his father: 'You run your business and I'll run mine. You supply the oil, and I'll refine it. The days are gone when one man can control every detail of a great oil company.'

The rivalry came to a head during the period of retrenchment. The Emperor clearly wanted to make the strategic decisions affecting the future of his empire. But he was neither an officer nor a director of Tidewater. When it came to disposing of the western marketing properties that were losing money, George wanted to be the one who masterminded the transaction. He angered his father by negotiating the terms of Tidewater's sale of its western properties, including several tankers, to Humble Oil, a subsidiary of Standard Oil of New Jersey. They were to pay over $300 million cash for five supertankers and the service stations in California, Oregon, Washington, Arizona, Nevada, Idaho and Hawaii. Unfortunately for George, the anti-trust division of the Justice Department thought that Jersey should open its own gas stations in the west if it wanted to enter the oil business. The government sued Tidewater and stopped the deal in its tracks.

Getty was quick to strike back in the power game with his son; he won his revenge when Phillips Petroleum, the largest shareholder of his antagonist Aminoil, decided to acquire the western properties. He negotiated the terms of the transaction – $309 million – in Rome at the Flora Hotel with Phillips's president, John Houchin. As a result, Houchin was to have personal difficulties with George because his 'father had cut off the negotiation process'.

Before Phillips's $309 million could pass to Getty, however, there would have to be another battle with the Justice Department. This time Getty's lawyers made Phillips put up a $5 million 'fight fee' that would be lost if the company backed off in any confrontation with the government. During the government's request for a court order to stop the transaction George 'beat the injunction by claiming from the stand that he would scrap all the service stations', according to Moses Lasky, Getty Oil's attorney. This claim, whether true or not, persuaded the judge to let the transaction go through, subject to a full trial of the case on its merits later on. It was a brilliant deal, because Getty Oil received the $309 million while getting rid of a heap of marginal refining and marketing properties. Moreover, Phillips agreed to purchase crude oil from Tidewater for ten years.

The giant sale, the largest in Getty's career, was a major step towards shrinking Getty's oil holding. He had given up his dream of a giant integrated company and from now on was to concentrate on exploration and production. Paul Louis Weiller felt that 'Paul was the first important oilman who saw that you should concentrate on exploration and oil reserves instead of marketing and refining.' A series of further asset sales improved Tidewater's profit statement. The cash from this retrenchment was also used to pay off Tidewater's debt, making the company far more financially stable.

Other cash available from the Phillips transaction was used to explore for more oil and increase the recoverable oil reserves. In 1966 alone Tidewater drilled 519 producing wells, a substantial increase over the 347 drilled in 1965. The introduction of the thermal recovery extracting process, by which steam is injected into a depleting oil reservoir, helped substantially to increase the life of the Kern River oilfield in California. The steam, injected into the wells, reduced the oil's viscosity, and it could be then driven up to the surface. At one time there were 300 huge steam generators with large furnaces helping to pump oil out of the California properties.

By 1966, as the Neutral Zone oilfields were reaching their peak and production was starting to level off, Getty must have realized that Tidewater held the most valuable crude oil in his empire. It now made sense to consider merging Getty Oil and Tidewater. While Getty Oil controlled by that time 72 per cent of Tidewater, Getty knew that any merger offer must be fair to the Tidewater minority shareholders or he would face stockholder suits again.

To accomplish the merger Getty needed to push up the market price of Tidewater. Suddenly, in the autumn of 1966, when Tidewater shares were selling at $55 a share, he arranged to have Tidewater offer to purchase in the market 1.1 million shares of its own stock at a price of $74 a share. In this way Tidewater reduced the number of shares left. This meant that now Getty Oil controlled a larger percentage of the remaining Tidewater shares, and since the assets underlying the shares were proportionately larger, the shares became more valuable. Getty was thus able to turn the cash he received from the sale of his western marketing and refining properties into an investment that would multiply his net worth yet again.

The offer of Getty Oil shares for Tidewater shares other than those already owned by Getty Oil reduced Getty's voting power in Getty Oil from 78.66 per cent to 62.24 per cent. His absolute power was not reduced, however. Under Delaware law the owner of 50.1 per cent of a corporation's voting shares can replace any officer or director, and thus influence any material decisions. J. Paul Getty and the Sarah Getty Trust would still own 12,573,789 shares of Getty Oil, worth at the 1967 peak about $1 billion for the first time in history. On 11 August 1967, the historic day when Getty mailed his merger plan to shareholders, he was worth $356 million in Getty Oil stock, while the trust was valued at $612 million.

All the time that Getty was increasing his fortune, his relations with his four sons were unravelling still further and threatening to cause him potentially serious legal and financial problems.

J. Ronald, the son of Fini, Getty's second wife, was half-German and at times inclined to arrogance. He was the largest of the four boys, 6 feet 4 inches tall, and sometimes sported a moustache like his father in his youth. His brief stay in what his mother described as an institution 'for difficult boys' as a teenager was followed by a degree from the University of Southern California in 1953, and six years later he was a success as Tidewater's marketing manager, making $40,000 a year. However, in 1958 Tidewater's then president, Dave Staples, aggravated by Ronald's interference in top management decisions, fired him and forced his resignation from the board. Staples felt that Ronald was 'causing too many problems'. Getty then appointed Ronald as head of Tidewater's Veedol motor fuel operation in Germany. 'Ronnie is just like me,' he told Staples. 'I see Ronnie in me.' Maybe he did. Ronnie became involved in a French lawsuit brought when Veedol enticed fifty employees away from a French lubricant company, LABO. LABO's owners brought an action in Paris against Ronald and several other Veedol officials but the sordid affair never received any publicity. It was hushed up in France and never mentioned in any Tidewater or Getty Oil financial reports. Ronald testified in 1980: 'I was a defendant, co-defendant in a case in France. The charge was taking employees away from a company . . . but there is an amnesty on that thing. In France it is illegal to hire away employees from another company. The company was fined, I was not fined.'

Ronald was the son who bore the emotional scar of being limited to a mere $3000 annual income from the Sarah Getty Trust. To compensate for this unfair treatment, which was the result of his father's bitter fighting over divorce and money with Fini, Sarah Getty had left Ronald $200,000 in 1941, and this had grown to a small fortune of $1.6 million by 1960 when it was handed over. At the time it made Ronald far wealthier than any of his brothers. What he did not know was that Sarah had meant to leave him a great deal more, a plan that was stillborn when Getty promised her that he would some day put Ronnie on an equal footing with the other boys. Ronald says, cynically and incorrectly, 'The only thing I inherited from my father was his love of animals.'

J. Paul Getty Jr, the son of Ann, the fourth wife, was considered the best-looking, most charming and most intelligent of the four sons. Yet he had a chequered time, leaving San Francisco State University and educating himself in literature, music and the arts. For some reason he too wanted to 'help my father', and, after a brief apprenticeship in California, he was picked to head Getty Oil's refining and marketing operation in Italy. With this offer Paul requested permission to change his name from Eugene Paul to J. Paul Jr. He was totally unsuited and untrained for the job, and he was bound to fail. Claus von Bulow, who was a friend of Paul's, described it as 'a demoralizing and destructive posting'. Paul's wife, Gail, the daughter of a federal judge in San Francisco, never thought 'business was his cup of tea. He was very artistic and never should have tried.' Paul felt that his father 'resented that he had to consider us in any way. He liked the semblance of paterfamilias as long as we were only seen and not heard.' From the moment he got to Rome, 'Paul simply never knew what he was supposed to be doing from one moment to the next,' says his oldest friend, Bill Newsom, who visited him there. By the mid-1960s Getty had fired his son as director of Getty Oil operations in Italy, blaming him for the $2 million annual loss. Paul Jr then turned to Rome's seductive *dolce vita* and was almost destroyed by it.

Gordon, the fourth son, was a gawky, unloved boy, ignored by his father. He grew up with an inner stubbornness, a confidence about his intellectual ability that seemed odd for such a vague, forgetful young man. But he was so overshadowed by his flamboyant brother Paul that he was the last son anyone would have expected to challenge his father's absolute authority over the family fortune. Gordon had neither George's executive ability, nor Paul's charm and brilliance, nor Ronald's arrogance. But he had a quiet intelligence and, as a youth, he kept largely to himself, concentrating on music. He cherished dreams of a career in opera. He also saw much of the family of Bill Newsom, his brother Paul's best friend, whose stable home was full of warmth and lively discussion. Gordon used to play Pedro, an Italian card game, with Newsom's father, a construction magnate and Democratic Party bigwig, whom he affectionately called 'Boss'. Once, Gordon lost $50,000 to him and, as he made out a cheque, he quipped, 'It's not good right now, but it will be.'

Like his brothers, Gordon seems to have wished to establish some connection with his father by working for him. He got into trouble on his first assignment in the Getty empire, as an aide in the Neutral Zone, when he refused to obey regulations and decided not to turn over two employees who were being sought by the local authorities. Placed under house arrest by the Emir for 'challenging his authority', he finally left the Neutral Zone for London, where, according to Everette Skarda, his father told him, 'I sent you out to be trained, not to be the director.' He worked briefly for a number of Getty divisions, mostly as a consultant, but never held down a permanent position.

Gordon never seems to have had enough money, or so he thought when he married Ann Gilbert, the tall, lively daughter of a Wheatland, California, nut and fruit farmer. Now that he was a husband, and soon to be a father, he needed an income. Until he was twenty-five he counted only on the $9000 he received from the Sarah Getty Trust. From then on he was due to receive a share of 20 per cent of the trust's income, but since his father was

conserving cash in the early 1960s and not paying a cash
dividend there was little income to come from the trust.
By now Gordon seems to have become absolutely fearless
as regards his father, and began to write him letters
insisting that Getty Oil Company should begin paying a
cash dividend.

'I believe Getty Oil is extremely well able to pay a
dividend,' he wrote on 9 January 1963. 'Failing this, you
as trustee are empowered to convert some or all of the
1934 Trust's assets from Getty Oil to income-yielding
securities.' Gordon was challenging his father's total
autonomy over the trust just as his father had challenged
George Getty I many decades earlier. If Getty Oil shares
were sold and replaced with a portfolio yielding 5 per
cent, Gordon calculated, his share of the income would
be $50,000 a year. In presenting this demand, he wrote:
'As I have practically no income at present, I would
appreciate whatever speed is possible in receiving your
permission on the above.' It was a demand his father
found outrageous. Getty had 'the same attitude that my
parents had. I never had any money myself, I was not
born with a silver spoon in my mouth, my father did not
believe that young people should have a secure income
whether they worked or whether they did not work and
that they should loll around in luxury.'

Getty dispatched George to offer Gordon a job in the
Getty organization at $750 a month, which represented
quite a saving on $50,000 annually. Gordon was not to be
pacified. He told George he could make more than $750
a month as a ski bum in Squaw Valley, a popular
California ski resort. Getty followed this approach with a
letter to Gordon on 7 February 1963, saying: 'I do not
think it is reasonable that the Board of Directors . . .
should abdicate and allow you to run the Company and
determine its dividend policy.' Back came a cable from

Gordon: 'Must seek legal declaration of rights.' A public fight in the courts between Getty and one of his sons was the one threat that always shook him.

The old man was angry, but he was also intimidated. His image would be stained by a lawsuit over which 'the world press would have quite a field day'. He was scared, too, that it might 'raise a question as to the validity of the trust'. Amazingly, he instructed his eldest son to capitulate to the demands of his fourth son. Getty Oil was to declare a cash dividend of 10 cents a share, a small amount really, but enough to assure Gordon his desired $50,000 annual income and with luck keep him out of court. This step was not exactly a disaster for Getty Senior, as the 10 cent a share dividend meant almost another $500,000 annually in his own pocket.

All the same, Gordon was going to have to pay a price. Getty was 'shocked and mad at Gordon trying to blackmail him', George noted at the time, and, 'doesn't ever want to see Gordon again.' There was to be other swift retribution. Gordon was to be removed from the list of potential successor trustees to the Sarah Getty Trust and banished from the museum's board of directors as well. He was prohibited, wrote George, from ever working for the Getty Oil interests 'as long as Father lives.' Last, but not least, it was codicil-writing time at Sutton Place again. On 15 January 1963 Getty struck out Gordon as an executor, leaving George, Ronald and Paul Jr. George and Paul Jr were now in line to get 20,000 shares of Getty Oil stock on his death. Ronald was to get only 10,000 shares. As for Gordon, he was to get not a single share of Getty Oil stock, but only the nominal sum of $500. The Emperor had banished his son, it seemed, but a few weeks later George slightly eased the ban on employment by making Gordon 'ineligible' for work unless he could be employed 'capably, competently and conscientiously'.

Gordon returned to the attack. He demanded 'redress, protection and vindication'. If he was not to be deemed a worthwhile son voluntarily, then the litigation over the trust would be instituted anyway. The Emperor backed down again. He denied charging Gordon with 'blackmail' and emphasized that George was wrong in saying he did not want to see him again. But he added: 'I would be very pleased if you would follow the example of your brothers and work for a living.'

Undaunted, Gordon refused to moderate his challenge. He wrote to his father on 17 August 1963 that he stood as a 'lone wolf' who 'had no interest in securing [his brothers'] cooperation, sympathy or even neutrality, as I considered these irrelevant.' He had another financial plan. He saw no reason why the court should not decide to grant approval for 'periodic partial distribution of the Trust corpus'. The very idea must have shaken Getty to his roots. Such a move would dissolve in part the body of wealth he had created, simply in the interest of making Gordon rich without his having to work for it.

Getty tried to pacify his son by telling him, 'Your musical compositions are very good,' but, with typical ambivalence, he could not resist adding, 'You are not sufficiently mature and do not now possess the qualities needed to function properly as a co-trustee of a trust the size and importance of the Sarah C. Getty Trust.' But he did make a concession. He hired Gordon as a consultant to analyse Tidewater's money-losing eastern properties. This role would place him in direct confrontation with George, so that the troubled relationship between father and son would now be transferred to the half-brothers.

George already had enough problems of insecurity with his father alone, Once, in Naples, other oil executives at a hotel meeting were startled to hear him ask, 'Father, is it all right with you if I go to Pompeii tomorrow?' Now he had to worry about Gordon challenging his authority, a serious matter, because Gordon blithely began criticizing Tidewater's senior management.

Gordon wrote an extraordinarily detailed and assiduous report, in effect chiding his father and George for holding on to money-losing properties. First and foremost, he advised that the Delaware refinery, then losing over $10 million a year. should be sold or mothballed. Feeling threatened, George decided that attack was the best defence and told his father in a letter on 17 June 1965 that Gordon would never become a 'well-rounded and seasoned business executive'. By August George was hitting below the belt. He brought up the old family gossip that Gordon bore a physical resemblance to Garret McEnerney II, a San Francisco lawyer whom his mother married many years after Gordon was born; this snide attack was totally untrue. No one ever seriously disputed the identity of Gordon's father.

Events moved steadily but firmly towards a crisis. When Gordon asked for a real job and George offered him the post of executive assistant to the general manager of the eastern division at a salary of $21,336, Gordon said he only wanted the job for ninety days. Then he decided to go to court in order to seek his reinstatement as a trustee. With some justification, he wrote to his father on 4 February 1966: 'I cannot imagine that you would have expected me to do anything else at this stage, or that you would have done anything else yourself in my position.' In fact, it had taken his father four fierce years of battle with his mother, from 1930 to 1934, to get control of the family oil interests.

When it came to the point, however, Gordon's lawsuit was not about his reinstatement as a trustee. Deciding that money was more important than position, he attacked the inviolability of the vast wealth represented by the Getty Oil shares in the Sarah Getty Trust. At the time practically 50 per cent of the shares of Getty Oil, worth $293 million, were owned by the trust, which had grown at a rate of 16 per cent compounded annually since 1934.

During the 1950s many American companies, instead of paying cash dividends to their shareholders, issued stock by way of dividends in order to preserve cash for new equipment or other expansion plans. Getty had also followed this practice, paying a 10 per cent stock dividend in 1953, 5 per cent in 1954, 5 per cent in 1955 and 5 per cent in 1957. Before the merger with Getty Oil, Tidewater followed the same programme. Getty's minority share-holders could either keep the stock dividends to enlarge their holdings or liquidate them for cash. Getty himself, and in the trust, kept all the additional shares issued as stock dividends. One day the shares in the trust, accumulated over many decades, would be divided among his grandchildren.

Here was Gordon, who was already being paid an income of $50,000 a year, demanding that all the dividends paid by Getty Oil to the trust should now be distributed to the income beneficiaries. It amounted to about 2,469,359 shares, or 31 per cent of the stock held by the trust, valued at $91 million. Gordon's share would amount to about $5 million. What he was in fact attempting was a partial liquidation of the corporate empire Getty had built in his mother's name. If Gordon succeeded, the trust would no longer have nearly a majority of the Getty Oil stock, as Getty had promised his mother. Worse still, the seventy-four-year-old oilman would have to pay a tax on the $72 million worth of stock he received, forcing him to sell part of the 30 per cent share in Getty Oil he held personally.

This potential disaster had to be avoided at all costs. To defend the trust and Getty, Lansing Hays chose a distinguished trial lawyer in San Francisco, Moses Lasky, a senior partner in the prominent firm of Brobeck, Phleger and Harrison, who had represented Tidewater in its struggles with the Justice Department over the sale of the western properties. A *magna cum laude* graduate of Colorado University at nineteen, Lasky was a brilliant, tenacious advocate in court.

Gordon chose his friend Bill Newsom as his attorney, along with James Martin MacInnies, Newsom's partner, plucky Stanford law graduates, who strongly believed that, under California law, Gordon should win the day. All they had to prove was that stock dividends, extra shares issued to shareholders in Getty Oil, should be given to Getty's children rather than kept in the trust.

Gordon intended to argue that his father had 'disregarded the working of his mother's trust fund by building up its capital value rather than distributing the income.' Lasky needed to prove that Sarah Getty wanted the Getty oil interests to remain in the trust for future generations of Gettys and not to be distributed as income. She had herself testified in 1940: 'We put money and notes in the trust for Paul to run the business. I wanted the property to stay in trust.' Lasky underlined this sentence in his brief.

While Gordon's case crawled through the courts, George's personal and professional lives were suffering rather severe setbacks. In the first place, his wife Gloria divorced him. Rather than be straightforward with his father about the situation, George cabled Sutton Place that the divorce was due to business travelling. His second blow came when Getty finally decided to merge Tidewater Oil Company, of which George was president, and Getty Oil Company, where Getty himself had the title. In 1967

George, then forty-three years old, running an integrated oil operation that was infinitely larger than Getty Oil, must have thought he would be named president of the combined concern, while his aged father took the graceful opportunity to become chairman. But Getty wanted no lame-duck status for himself. The Emperor remained president and kept George under his thumb as executive vice-president and chief operating officer. 'George was terribly disappointed, and bridled constantly at the way his father ran roughshod over him,' says Stuart Evey, who was close to the younger man.

In fact, George posed no threat to the status of the trust, or to his father's personal tax position. The big danger was the suit brought by Gordon, who was no longer even considered as part of the dynasty. Getty had no intention of handing Gordon $5 million for hardly a day's work and giving the tax collector $25 million.

Lasky claims that, if he had been representing Gordon instead of his father, he could have won the case. Perhaps he had in mind a 1941 amendment to the 1934 trust which, he says, was 'typed on a slip of yellow paper and clipped to the typewritten form'. Mysteriously, it was never produced at the trial. No one can explain its omission, but in any case Lasky says it would not have swung the case in Gordon's favour. Gordon himself did not know of its existence until he opened one of his father's safe deposit boxes in 1976 after his death. Bill Newsom says that the lack of the document 'was the only way we could lose the case. It was the crucial point of the case.'

The amendment was dated April 1941, that is after the courts had confirmed the irrevocability of the trust but shortly before Sarah died, and it turned the purpose of the Sarah Getty Trust upside down. Put in simple terms, it said that, in any dispute over the money in the trust, the income beneficiaries – that included Gordon – were to take precedence over the ultimate recipients, that is, the grandchildren of J. Paul Getty. Its actual wording was:

The parties hereto do hereby declare that the primary purpose and intent in creating the trust . . . was and is to provide for the income beneficiaries of said trust, and the rights and interests of all remaindermen are and shall always be subordinate and incidental to that purpose, and the provisions of said trust shall always be liberally construed in the interests of and for the benefit of the income beneficiaries of said trust.

The existence of this amendment is evidence that Getty had pulled a final fast one on his infirm and totally deaf mother just before she died, for it gave him the right to take precedence over her grandchildren and great-grandchildren, whom the trust was supposed to protect.

Another document not produced at the trial was a letter from David Hecht, Getty's lawyer until 1959, suggesting that 'stock dividends regularly declared and paid in lieu of cash dividends shall be income'. Either the existence of this letter was unknown to Getty's lawyers or they ignored it. Indeed, when the trial finally took place, Lasky took pains to destroy any notion that stock dividends were the equivalent of cash dividends. 'A stock dividend is nothing but a piece of corporate book-keeping,' he insisted. 'It creates nothing, changes nothing, takes nothing from the corporation declaring it, gives nothing to the stockholder he did not already have.'

While the bitter court battle was proceeding, Ann Getty, Gordon's wife, eager to keep the lines of communication open with the Emperor who controlled her purse strings, called him at Sutton Place and complained, 'Your lawyer is killing my husband.' The next afternoon, in San Francisco, Moses Lasky's phone rang. It was his client calling from England. The hoarse voice croaked, 'Keep killing my son.'

The battle royal over the dynasty ended on 30 October 1970. Judge Charles Peery of the California Superior Court ruled that plaintiff Gordon Peter Getty 'is entitled to have and recover nothing whatever from the trust estate or from J. Paul Getty as trustee or individually.' The Sarah Getty Trust, just as its sole trustee wished, was alive and well, 'primarily to the end of preserving and enhancing a capital and corpus for distribution to his descendants.' J. Paul Getty had preserved the trust and, with it, his status as one of the world's certified billionaires.

17
Three Tragedies

Pity the rich. In terms of living they are beggars.
J. Paul Getty III

While Getty could win lawsuits which involved his fortune, he was unable to prevent three terrible tragedies from striking the Getty family in the early 1970s. A daughter-in-law died of an overdose of heroin, a son probably committed suicide in the most extraordinary circumstances, and a grandson was kidnapped and mutilated. How far these tragedies were in some way the outcome of Getty's remoteness from his sons must be a matter of conjecture. He had the same ambivalent relationship with them as he did with his own father and mother. He loved them after a fashion, in his own quirky, inconsistent way, but he also sent them messages of contempt and indifference. Coupled with his aloofness, one also has to take into account their own professional failure when compared with their awesome father whom, at different times and in different ways, they sought to impress. It is, perhaps, not surprising that a son and a grandson became victims of addiction to alcohol and drugs, and that the pressure of trying to become heir apparent might have caused another son to court disaster with a combination of alcohol and barbiturates.

Getty himself must have understood some of this. 'There are people,' he said, 'who have been destroyed, physically and morally, by their wealth. The same people, born poor, would probably have become alcoholics or thieves.'

* * *

His third son and namesake, J. Paul Jr, was probably the lover of more exotic women than even his five-times-wed father. The only problem was that he was wilder too. With his long hair and full beard, he was featured in 'scandal magazines across the globe', his father complained, 'in a tie-dyed velvet outfit that would make any genuine hippie green with envy.'

Paul Jr always drank too much as a young man, partly from a youthful ambition to be 'a character in a Fitzgerald novel,' according to his mother. She tried to warn his future parents-in-law about him before he married his childhood sweetheart, Gail Harris, the daughter of Federal Judge George Harris, in 1956. They had four children together – Jean Paul III, Aileen, Mark and Ariadne. But the seductions of the fast life in Rome during the early 1960s were appealing and helped to shatter this youthful marriage in 1964.

More than any Getty son, Paul Jr was 'a great success with those social friends of the father's in England, in France and in Italy, whom he admired,' according to Claus von Bulow, his close friend. 'He had charm, he had conversation, and he had sex appeal.' Through von Bulow, Paul Jr met a woman who was to be the love of his life.

Talitha Pol was a shimmering beauty with red-gold hair, almond-shaped eyes and white alabaster skin. She was intoxicating, a 'wild bird' who designed her own clothes in the most startling colours and was the 'feyest, most wonderful person' he had met. The daughter of a Dutch painter, William Pol, Talitha was brought up on the exotic island of Bali. During the Second World War she and her mother were imprisoned by the Japanese in a prison camp; her father was held in another. Her mother died soon after the war ended, while looking for her husband. William Pol then married Poppet John, the daughter of the painter Augustus John.

By all accounts Talitha was full of high spirits and humour, but tinged with a sense of insecurity and slight melancholy. Zeri found her to be a *femme fatale* at the age of fifteen, a creature who made Getty nervous. She was far more exotic than any of the women Getty had known, and her death was a tragedy which cemented the fate of Paul Jr. Getty had never let any woman except his mother undo him, but Paul Jr and then George lost control.

When they married on 10 December 1966 Paul was living on his growing income from the Sarah Getty Trust and Talitha was an actress who had small parts in some of the films being made in Rome. At the wedding Talitha wore a velvet miniskirt trimmed with white mink. Jack Sunderland, president of Aminoil, was visiting Sutton Place at the time and says that Getty 'didn't even know Paul was marrying Talitha Pol. I asked him how many grandchildren he had. He didn't know.' In Rome Paul Jr and Talitha were part of the beautiful people depicted in Fellini's *La Dolce Vita*; they entertained in a living room that contained a giant birdcage, two elephant chairs from India and a shrine to Buddha with incense permanently burning. On the wall of one room was a photograph of Talitha, her back to the camera, holding up her skirts to reveal all her charms.

The last years of the 1960s were for Paul a far wilder equivalent of his father's freewheeling on the Continent in the 1930s and the 1950s. His friendship with Mick Jagger and the Rolling Stones led to some splendidly hedonistic days in a Marrakesh villa, photographed in loving decadence by Horst for *Vogue* magazine. They sailed with the German steel heir, Arndt Krupp, on his yacht through the Mediterranean, and Talitha would appear in the mornings in a skintight diving costume with a knife strapped to her leg. In Rome they attended the notoriously wild parties thrown by Dada Ruspoli, where they mixed with film personalities, the titled nobility and the ersatz jet set known today in New York as 'Eurotrash'. In this social milieu, it is easy to understand how Paul Jr, who began as a heavy drinker, might become susceptible to the drugs he saw everyone taking. Tony Sanchez, who regularly obtained drugs for the Stones, described him as 'so chronically addicted to smack that he was snorting a grain a day without managing to achieve any notable effect.' He did not have the self-control of his friend, Mick Jagger, who said, 'It's all right letting yourself go, as long as you can get yourself back.'

It was apparent how deeply the hippie, bohemian life had taken root when Paul and Talitha named their baby son Tara Gabriel Galaxy Gramaphone Getty – a name that infuriated the elder Getty at Sutton Place. He also became enraged that Talitha was pictured in the French edition of *Vogue* with nothing on but a small pantherskin costume. 'This is not worthy of a Getty,' he complained. 'I have to insist that if such a thing occurs again I shall disinherit my son. I will not allow that my name be dragged in the mud.' In response Talitha told the press, 'Who would imagine that he is so prudish?'

After a while, however, Talitha wanted to break out of the Rome drug scene and moved to London where she began to take up with other lovers. According to Gail, Paul 'was horrified by the news she was going to divorce him.' Talitha returned to Rome in July 1971 at Paul's request, to discuss a possible reconciliation. On the morning of 11 July she was found dying, apparently from an overdose of barbiturates. No one knows what had gone on the previous night; it was not until six months later that the police decided that Talitha had died from a massive overdose of heroin.

Immediately following Talitha's death Paul had gone to see Gail, who was staying at Palo, his father's villa on the Mediterranean. 'I called father to ask his permission to stay there, and he hemmed and hawed and said, "Well, I dont want any wild hippie parties there." As if I was going to have any wild hippie parties with Talitha just dead a couple of days. And I lost my temper at that one. I lifted the phone and looked at it and I just said, "You bastard," and rang off.'

Back at Sutton Place the old man responded in his customary style. On 29 July 1971 he drew up the four-teenth codicil to his will since September 1958. This one dealt with Paul Jr rather than with one of the women. Paul Jr joined his brother Gordon in limbo, being removed as an executor and no longer receiving any shares of Getty Oil upon his father's death, but only the insulting bequest of $500.

Some seven months later, in February 1972, Paul Jr moved to London, apparently because the police in Italy wished to question him about Talitha's death. His ex-wife, Gail, remembers that the Rome newspapers were full of stories about Paul being wanted for questioning, and of a possible indictment for negligent homicide.

Paul Jr, now living at his London home on Cheyne Walk, sought reinstatement as the favourite son – but his father continued to rebuff him. He told Bramlett, 'I'll have nothing to do with him until he quits using those drugs. Then I'll take him back into my affection.' Getty's fear of drugs and his antipathy for anyone using them were so strong that he never saw Paul Jr again.

Many years later, in the early 1980s, Paul Jr was to treat his great friend, Claus von Bulow, with far greater generosity at a time of grave threat to von Bulow's freedom. Paul Jr lent von Bulow $1.5 million to defend himself against charges that he had attempted to kill his socialite wife, Sunny, through lethal injections of insulin. Paul Jr felt that Claus, entirely innocent like himself, was the victim of an ambitious prosecutor trying to make a name. Roughly $1.5 million intended for von Bulow was first deposited in a nominee account at Macfarlanes, Paul Jr's English solicitors, and then it was redeposited in the Rhode Island Hospital Trust National Bank in Providence, where it was available to von Bulow.

The identity of von Bulow's financial angel was a well-kept secret, apparently even to his lawyers. In 1984 his defence counsel, Harvard law professor Alan Dershowitz, clearly obtained a reversal of von Bulow's conviction, although von Bulow remained under indictment. An application was made to the Rhode Island court to reduce his $1 million bail on the ground that it was a financial hardship to the group of anonymous donors who put it up. All but $10,000 of the $1 million remaining from the money von Bulow had originally received from Getty was subsequently returned to von Bulow and used for the second trial.

If Paul Jr had been the favourite son personally, perhaps George, the eldest and heir apparent, was the most honoured. He was executive vice-president of Getty Oil and had two votes as his father's executor. Moreover, he was a respected business leader in southern California, a director of the giant Bank of America, the Douglas Aircraft Company, a chief fundraiser for the Los Angeles Philharmonic and a successful horse breeder. Still, he was also miserable at never receiving praise for his business success. Rather, from Sutton Place Getty sent to Los Angeles a stream of sarcastic, highly critical memos designed to remind George that his father was watching constantly and without admiration. The abuse was often sandwiched between the greeting 'Dearest George' and the parting 'Love Father'.

But George must have feared for his job on 31 December 1968, when his father warned that 'if a company's management cannot equal the industry average [of profits] I think the stockholders should replace it.' Of course, Getty was the largest stockholder – as well as the president – and George would get no solace from his father's musing a few days later: 'It would seem that if GOC cannot do better than predicted by its present management, I should ask Dr Hammer [chairman of Occidental Petroleum] or Mr Robert Anderson [chairman of Atlantic Richfield] to replace me.'

That was an idle threat compared to the elder Getty's frustration at George's inability to find oil in Algeria, Pakistan or the Spanish Sahara. From 1969 on he insisted on approving every exploratory oil well that cost more than $450,000. 'I would not approve drilling any well estimated to cost over $500,000 unless it had a chance of one in four and if it was estimated to cost $1 million or more it must have a chance of one in three.'

Getty threatened even more severe action if George did not reduce the large losses at one of their subsidiaries, Nuclear Fuel Services. Get the red ink down to $3 million or 'I will drop an H bomb on it,' he wrote.

After the Duchess of Argyll returned to England from a trip around the USA on which she had been escorted by George she teased Getty about her friendship with his son. Getty replied, 'But don't you find George so boring?' Dutiful George, ever trying to impress everyone, 'was old when he was young. So much responsibility,' says Gail Getty, Paul Jr's ex-wife. The Duchess of Argyll claims that George told her how miserable he was about his father and 'how deeply he hated him'.

In 1951 George married Gloria Gordon, a dark-haired, pretty woman from Denver, who reminded some members of her family of Elizabeth Taylor. Needless to say, Getty was not at the wedding. Indeed, he did not attend the wedding of any of his sons, but then he had not invited his parents to any of his five weddings.

George and Gloria had three daughters – Anne, Claire and Caroline – born during the 1950s. The great pressure for George to succeed in the Getty Oil business, however, apparently caused him to concentrate on his business at the cost of his personal relationships in the same way, but to a lesser extent, as his father had. When, after sixteen years of marriage, Gloria filed for a divorce in 1967, she charged that her husband was 'aloof, cold, indifferent and insulting'. She told the Los Angeles Superior Court Commissioner that 'he refused to let me go with him – and when he was home, he stayed to himself.'

Despite his executive position and name, George was not a great ladies' man like Paul Jr. Yet he was quite a prominent catch for Jacqueline Riordan, the beautiful, raven-haired widow of Michael Riordan, a riproaring Irish-American mutual fund salesman who had helped start Equity Funding Corporation, a mutual fund organization that sold insurance and became in the late 1950s one of Wall Street's hottest stocks, and then one of its greatest scandals when it was discovered that a large portion of its insurance policies were bogus. During one of Los Angeles's torrential rainstorms, a mudslide overran the Riordans' home and trapped Michael in bed. Luckily for Jackie, she was in the bathroom. She heard 'an enormous rushing sound and tried to pull him out of a massive gob of wet mud', but her multimillionaire husband was drowned in it.

Jackie met George Getty II at a Los Angeles social function and found him a 'dutiful, caring man'. George liked her not only for her lively beauty but also because 'she was richer than he was', according to Stuart Evey, George's closest colleague. George had become the owner of thoroughbred horses and taught Jackie the fascinating sport of kings. He contrived to be stolid and romantic at the same time. Unlike his father, he could inscribe adolescent, sentimental lines like 'The most beautiful garden is the garden of your heart.' George and Jackie were married in May 1971. He was happy to move into her Bel Air home, which was even larger than his fifteen-room French chateau in Beverly Hills. There was plenty of room for her two sons and daughter as well as his three daughters.

Von Bulow felt that 'George was as conventional an American executive as the father was an unconventional one . . . If Paul had taken as dispassionate a look at his sons as he did on his material assets, he would have cast each of them in different roles. George would have been promoted as an excellent hail-fellow-well-met politician. A good Governor of California, a Senator, and a prominent racehorse owner.' Instead, Getty often treated the heir apparent with cold disdain and humiliated him in front of Getty Oil personnel and his friends. Once George arrived for his regular business discussion with his father and was kept waiting for eight days for an audience. Another time George invited three couples from Los Angeles to Sutton Place to meet his father and the old man refused even to come down for a meal. George always felt that his father blamed him for everything that went wrong, whether it was the fall of Getty Oil shares in the stock market or a road accident that held up a shipment of Getty's French furniture for his museum. George told his friends he felt like the 'vice-president in charge of failure', while his father was the 'president in charge of success'. When the London *Daily Mail* asked

Getty to describe his son in 1972, all he found to say was that 'he lives in Los Angeles'.

It is hardly surprising that George was a man of extreme moods. He might try to prime himself with the strains of a John Philip Sousa march before leaving for work or, in more sombre mood, visit a graveyard. He could be an impulsive drinker, and it was revealed after his death that he relieved his feelings on occasions by stabbing himself in the hand with a letter opener. Many of the top executives at Getty Oil headquarters found him 'pretty tightly strung', especially after long transatlantic phone conversations with his father.

On the afternoon of 5 June 1973 George instructed Harold Berg, the head of exploration and production, to bid up to $100 million, a surprisingly large figure, in the forthcoming sale of an offshore lease in the Gulf of Mexico, and then drove out along Wilshire Boulevard, past the tall office buildings that had replaced the original orange groves, to his elegant house in Bel Air, the private enclave of Beverly Hills' wealthiest business and movie magnates.

His marriage to Jackie had been turning sour for some time – which added to his misery. Because of a 'personality conflict' between Jackie and her mother-in-law, George had not seen his mother since November 1971, according to a report by Thomas T. Noguchi, the chief medical examiner and coroner in Los Angeles, and Ralph M. Bailey, chief of the investigations division. In addition, relations between Jackie and George's three daughters had become so troubled that Anne, Claire and Caroline had moved out of his home to live with their mother.

'All I know is that he was fighting with her. I know he was goddamned unhappy with her and with his father and everything else,' says Stuart Evey, who knew the family well. George's relationship with Jackie 'became almost a combat situation under certain circumstances.' To help Jackie get into horseracing, George was selling her an interest in his horses, according to Evey. When the horses ran badly, however, 'she'd needle him about [them] . . . She overpowered him,' explains Evey, because for one thing George was worth only $300,000–$400,000 while Jackie had $30 million from the sale of her late husband's Equity Funding shares.

While George confided in Evey about his problems at home, he did not express his feelings about his father. However, George had been worried about 'maintaining his physical strength and his position in the business world', according to the report by Noguchi and Bailey. He was drinking more than usual and taking medication in order to sleep.

At dinner on the night of 5 June George, already 'irritable and irascible', drank several beers and an entire bottle of wine. After supper, while Jackie was walking the dogs, George appeared on the balcony with a shotgun which he kept for protection and fired one shot in to the air, according to Evey.

Jackie managed to hide the gun, but George ran into the kitchen and grabbed a barbecue knife, shouting that he 'was going to kill himself', according to the report. He drew the knife across his upper abdomen and inflicted a 'small superficial wound' about three eighths of an inch deep, drawing blood but not puncturing any vital organs.

Noticing blood on his stomach, Jackie telephoned Dr Kendrick Smith, the Getty Oil Company doctor, Evey and the Bel Air Patrol, the private police force which guards the enclave of the wealthy from robbery. Meanwhile, George, losing control, talked about his 'still being strong, powerful, masculine and able to bear pain'. He also spoke of death and suicide, stating he 'would use a gun or knife because taking pills was weak and cowardly'.

At 12.30 A.M. 6 June 1973, the Bel Air police arrived. George panicked and ran upstairs, locking himself in the bedroom. The bedroom light went on for a short time during which period he probably 'ingested the lethal dose of pills', later found to include Valmil, Miltown, Librium, Nembutal and Phenobarbitol.

From behind the closed bedroom door George yelled he 'had a gun and would kill anyone who came near the door'. He threatened to shoot six or seven people. Then 'his talk turned to moaning and then to silence'.

The Bel Air police were informed that George was sleeping off an alcoholic binge, according to Evey. Evey could hear 'a deep, intense rhythmic snoring' inside the bedroom. He could not open the door so he kicked it in and found the executive vice-president of Getty Oil, the heir apparent, in his jockey undershorts, drenched in blood from the self-inflicted wound.

At 1.02 A.M. an emergency call was made to the Los Angeles Fire Department for an ambulance. George was taken to the Queen of Angels Hospital in downtown Los Angeles where Dr Smith, the Getty Oil doctor, had visiting privileges, rather than the far closer UCLA Medical Center. Both Dr Smith and Evey wanted to avoid unpleasant publicity about George 'passing out from an alcoholic binge and a fight with his wife.' Neither man had knowledge at the time that George had swallowed a heavy dose of barbiturates.

George was admitted to the hospital at 2.40 A.M. where his illness 'was diagnosed . . . as a possible concussion.' Someone gave George's name as George S. Davis and his address as Evey's home in North Hollywood. The hospital was told that George had been 'barbecuing at his poolside earlier in the evening . . . and fell to the pool decking, suffering contusions and abrasions to his hands and body.'

It was not until the next day, when it was too late, that empty pill bottles were found in the laundry hamper at George's home. At the time he was admitted to hospital it appeared that he had simply lost control of himself through too much drink and needed to sleep it off.

Evey, who had accompanied his friend in the ambulance, stretched out on the floor of George's hospital room. He was awakened with a start at about 4 A.M. to find nurses and doctors 'running all over the place, because George had momentarily stopped breathing. He had gone into a dangerous coma.'

George's mother, Jeanette, and his stepfather, William Jones, were called to the hospital. A week earlier Jeanette had had a letter from George, similar to many other depressing notes she had received, saying he would 'see her in the hereafter, and happier life', according to the coroner's report.

Later that morning Evey went to his office and 'pretended that nothing had happened'. He did not dare tell a soul that George's life lay in the balance. When he phoned George's father in London he found him at the Duchess of Argyll's house in Grosvenor Place. 'I lied to him,' Evey said. 'I told him George had suffered a slight stroke and was unconscious in the hospital.'

At 2.30 P.M. the doctors called Evey to tell him that George had died. Once again Evey called Getty at the Duchess of Argyll's home in London to give him the traumatic news. For half an hour the eighty-one-year-old oilman stared dumbly into space. He would not accept the offer of a drink from the Duchess, but she was 'not surprised when suddenly the brief period of mourning was over. Paul collected himself, and coolly began figuring out who was to replace George and the management changes that were needed.' The next morning he had an appointment with Spencer Samuels, son of the art dealer, the late Mitchell Samuels, who drove down to Sutton Place with him. Samuels was amazed that 'Paul never mentioned George's death once'.

Later that day Jeannette Constable-Maxwell also drove down to Sutton Place. She walked with Getty to the chapel and they read the service to each other in memory of George. 'He had tears in his eyes as he read the lines from the prayer book,' says Jeannette. 'He had hoped George was taking things over, even if he wasn't up to it. Having to take things back into his own hands made him look tired for a moment. Then he had a thought. It was 10 A.M. in New York, and the stock market would be opening.'

George Getty II was buried on 9 June. At first, investigators 'found no evidence of marital, physical or financial problems' in his death and Coroner Thomas Noguchi disclosed that he was not drunk. The alcohol content in his blood was 0.06 per cent, less than the 1 per cent necessary for proof of drunkenness. The truth as revealed in the complete autopsy was stranger and worse than anyone had imagined. His body was found in a state of 'acute intoxification', not from alcohol, but as a result of the 'ingestion of multiple drugs and ethanol'. There was a detailed description of 'small reddish-blue and purplish bruises . . . along the backs of both hands and along the forearms and inner aspects of both upperarms.' He had been lacerating his hands with his letter opener. The coroner found 'multiple needle marks on both arms'. Veins in the region of the elbow 'showed evidence of penetration of a needle'. The stab wound on the evening of 5 June was a symbol of the damage George had been secretly inflicting on himself over a long period.

The conclusion of the coroner's report read as follows: 'From the evidence, Mr Getty's death was the result of a slowly increasing personal stress which was triggered into an acute reaction by alcoholic intoxication and which manifested itself in violent and self-destructive behavior. We believe Mr Getty took the pills in the brief time he was in the bathroom, that the pills were ingested on impulse, possibly motivated by feelings of guilt, shame and anger associated with concern about masculinity.' After interviewing Evey, Jackie, her son Michael, and George's mother, Noguchi and Bailey added: 'If death occurs as the result of behavior of high lethal potentiality, it is appropriately deemed suicidal.'

On 1 June, five days before his death, George had composed, in his own hand, a last will and testament which he mailed to C. Lansing Hays Jr, the Getty Oil lawyer in New York. 'I am of sound mind and body,' he declared, 'and I write and set forth this last will and testament of mine under no force or undue influence whatsoever.' He left 37.5 per cent of his estate to his wife Jackie, 62.5 per cent to his three daughters, and $10,000 to his father, J. Paul Getty. Was this bizarre bequest – a mere token, of course, in relation to the Getty millions – an expression of his anger at the way he had been treated all his life? Evey suggests that it was possibly his way of saying 'to Mr Getty, I guess the most important thing in your life is money. So I'll just try to help you along. He used to send the old man a little money from time to time – say $100 on Christmas and his birthday.'

According to Evey, 'Mr Getty did not convey to George how much he did admire him. George never got credit for making Getty Oil what it was.' Jackie believes, 'He'd still be alive today if it were not for his father. George was in awe of him, but was more in the image of his grandfather [George I]. He never did an incorrect thing in his life.'

Getty considered in his autobiography that the alcohol and barbiturates that George had been taking might have been due to the fact that 'he strove too hard to live up to the images of his grandfather and me'. Norris Bramlett, according to Getty, wrote a detailed report which concluded: 'I am convinced beyond any doubt that it was an accident.' William Hickey of the London *Daily Express* wrote that the tragic death of George 'has had a poignant effect on the family. It has drawn them closer together than they have been for years.' Much to the dismay of the Duchess of Argyll, Getty placed a portrait of George in a purple velvet frame in a prominent location in Sutton Place. It was the first picture he had ever displayed of George and she considered it an example of 'Methodist hypocrisy'.

On 14 June, eight days after George's death, Getty drew up his sixteenth codicil. This one was necessary. George was gone. The executors now were Ronald and the Bank of America.

If J. Paul Getty is remembered for anything in addition to the pay telephone he installed in Sutton Place and the art museum he established in Malibu, it must be for the vicious mutilation of his grandson by a gang of Calabrian kidnappers in 1973. The kidnapping of Paul III, the most bizarre incident in the history of the Getty family, was also the hideous climax to a record of parental alienation that began in Victorian times with the clash between Getty and his strict, self-made father, George I. The son protested his love and devotion even as he opposed his father and strove to overcome him. In his turn, he was equally aloof from his own sons, and the pattern was repeated in the next generation as Paul Jr, after his marriage to Talitha, paid little if any attention to any of his children. After Talitha's death Paul III told the magazine *Rolling Stone* that he and his father communicated 'through postcards and strange telegrams'.

'Little Paul adored his father and wanted to emulate his incredibly bizarre lifestyle. It was his way of trying to have a relationship with him,' according to his mother Gail.

The boy, born in 1956, the year before his grandfather was named the 'richest American', was the wildest of the three Pauls. He loved to drive his Harley-Davidson motorcycle at dangerous speeds through Rome's piazzas. He was unruly at school and was, according to his version, ousted from eight of them by the time he was sixteen years old. (Gail says he was asked to leave only one school.) He spent more time in his favourite Rome night clubs – Scarabocchio, Gattopardo, Tree Tops – than in school or, for that matter, at home. He enjoyed painting, which he did at the studio of his friend Marcello Crisi, and he often stayed with a beautiful, vague German girl, Martine Zacher. (He was later to marry her and have a child of his own, Paul Balthazar.)

By the summer of 1973 his mother Gail 'could no longer control him at all'. Bill Newsom, his godfather, found him 'a delightful boy who simply drifted away into trouble'. On one occasion he was wrongly arrested during a leftist demonstration. Despite protesting his innocence, he was thrown into one of Rome's most violent prisons where, according to his godfather, 'he was horribly beaten and sexually attacked'.

It is not surprising that someone, sooner or later, would have the idea of kidnapping the grandson of the world's richest man, who frequented the same Rome night clubs as Mafiosi hoodlums and drug dealers, whores and hangers-on. Between 1960 and 1973 there had been 320

kidnappings of wealthy children in Italy; it had become
almost a national sport. But the Getty kidnapping in July
1973 was to be a record breaker.

The essential narrative of the kidnapping that follows
is the result of an extensive investigation by examining
magistrates in the region of Lagonegro where Paul III
was finally released in December 1973. In addition, the
author has interviewed several of the participants.

It was about 3 A.M. on the morning of 10 July 1973,
just after Paul III had left a group of friends in the
Piazza Navona, where the resident hippies were joined by
jewellery and rug merchants selling their goods to tourists.
He was walking down a dimly lit street near the Piazza
Farnese when, according to his own account, several
men jumped out of a white car and 'overwhelmed me'.
Blindfolded and stunned, the boy was driven 240 miles
south to the bleak, stony mountains of Calabria, a primitive, stricken land, far from the glamour and excitement
of Rome's night life. For days that turned into weeks
Getty's grandson was forced from hiding place to hiding
place, caves, huts, farmhouses, sometimes simple sheep-
folds. He told *Rolling Stone* that he 'walked along narrow
paths, winding crevices, and steep cliffs, woods, hills,
open plains, and olive groves, and once in a while, made
short trips by car, with his eyes blindfolded, through
roads that seemed to him disconnected and twisty, from
one hiding place to the next.'

Paul's disappearance came five weeks after the death
of his uncle and almost exactly two years after the death
of his stepmother, Talitha Pol.

Two days later, in the afternoon of 12 July, his mother,
who had an apartment in the Trastevere quarter of Rome,
received a call demanding an unspecified amount of
money in return for her son. She told the caller her
resources were almost nil – indeed, she was finding it

difficult to get enough money to send the children to school. She 'had no doubt, none,' that the kidnapping was real. She fainted after getting the phone call; when she recovered she immediately called the police.

Right from the start there was doubt that the kidnapping was real. The Rome daily newspaper, *Il Messagero*, carried the headline 'Kidnap or Joke?'. The police at first felt the kidnapping was a hoax by Paul III, says Gail now.

The first word from Paul III came in the form of letters to his friends Martine Zacher and Marcello Crisi on 16 and 17 July, but which were meant for his mother. In them, he 'pleaded with her to show the letters to his father and grandfather' in order to get the money. Otherwise, Paul warned, he would be 'treated badly'. The kidnappers had threatened to cut off one of his fingers.

The first demand for money came on 18 July. The kidnappers called and asked for 300 million lira ($500,000). The Rome police felt that the moderate ransom demand was the work of amateurs and turned up evidence that Paul III had been socializing with known criminals. They were also suspicious for two other reasons. They told the press that Martine Zacher had claimed that Paul III had spoken about arranging his own kidnapping 'to raise money'. Also, Paul III had frequented a cinema that had been showing *Travels with My Aunt*, a film that includes a fake kidnapping in which the victim's finger is sent through the post as proof. (Much later the prosecutor's intensive examination of the case would point out that the Rome police were responsible for promoting the theory of the kidnapping as a hoax, which gave it a life of its own going beyond rumour.)

Gail, however, was convinced that her son was in

danger. She immediately sought help from both her ex-husband and the boy's grandfather. In the safety of Sutton Place, Getty stood tough. 'I have fourteen other grandchildren and if I pay one penny ransom, I'll have fourteen kidnapped grandchildren,' he told the press. He was suspicious about his grandson, because of the boy's radical politics, his escapade into Rome's underworld and, especially, his use of drugs. 'No Getty should be a hippie or a drug addict,' he told his friends. This eldest grandson had antagonized Getty by writing 'I am a refugee from a Rolls-Royce. I am an escapee from the credit card. I eat one meal a day. And life is a banquet. The rich are the poor people of this earth. They are a suffering minority whose malnutrition is of the spirit. Pity the rich. In terms of living they are beggars.' Getty's suspicions were heightened because Gail's boyfriend at the time was Louis della Ratta, a street-smart Italian-American who managed a restaurant in Rome. Gail always denied she or della Ratta had any part in the kidnapping. Nor did the Italian police find any proof of it. However, as the days progressed, Getty's doubts about the genuineness of the kidnapping were fuelled by reports that Paul III had been seen in Rome, walking around free when he was supposed to be in the hands of his captors.

During the six months that Paul was in captivity Getty tried to avoid responsibility for the event. He refused to talk to Gail, or to Paul Jr, preferring to work through intermediaries. He became even more afraid of being kidnapped himself. He began to drive everywhere with an armed car in front and another to the rear of his Cadillac. He instructed his aides to find out if the kidnapping was real; he was determined not to be the victim of a hoax by members of his own family.

Paul Jr, now living in London and 'in terribly bad

shape', according to Gail, also believed that the kidnapping was a hoax. 'I don't believe it. I didn't trust anything. Paul's life was so strange. . . . Other people I spoke to doubted it too,' he says now. However, he kept in touch with his father through various people, including Getty's secretary, Barbara Wallace, and later his negotiator with the kidnappers, J. Fletcher Chace.

On 24 July the kidnappers, or those posing as kidnappers, substantially increased their ransom demand to 10,000 million lira ($16.7 million). The police were confused by the dramatic change in the amount and remained unconvinced that the kidnapping was real. By the end of July the kidnappers themselves may have been confused by the conflicting signals being sent by members of the Getty family. At Sutton Place the grandfather was still adamant in his refusal to pay. However, on 31 July, Gail Getty's Rome attorney, Giovanni Iacovoni, told a caller that the distraught mother would pay 200 million lira ($330,000). The voice on the phone repeated the demand for 10,000 million lira.

A week later, on 7 August, another caller asked for 400 million lira ($666,000). Before paying, Iacovoni asked for proof that Paul III was still alive. There was no answer to this reasonable request.

By 15 August Paul III had been missing for more than a month and Getty decided to send J. Fletcher Chace, the manager of the Neutral Zone oil operation, to Rome to investigate and 'to negotiate . . . with the kidnappers'. When the kidnappers called on 18 August, Iacovoni told them that Chace was 'ready to pay a sum of about 53 to 54 million lira' ($90,000). A day later Iacovoni received a letter from Paul III which had been sent on 6 August in which he pleaded for 3000 million lira ($5 million), 'otherwise he would be killed'. Unfortunately, the way the sale of a work of art or an oil lease can be negotiated from the drawing room of a British manor house is not applicable to the cold-blooded negotiations as understood by violent criminals in Calabria. Chace's offer was rejected out of hand on 20 August. An agitated, gravelly voice told Iacovoni in 'vulgar, threatening' terms that the sum now demanded was 3000 million lira. Otherwise 'he would mail a leg or an arm of the boy and later his body.'

Chace decided to play detective; he was determined to bring the boy back himself. In the course of his investigation he met a young Italian, known as Bruno, in the Piazza Venezia on 28 August. Bruno said that he had been involved with young Paul and others in simulating the kidnapping and offered to take Chace, for the price of $500, to Paul's hiding place in an old mill near Monte Cassino, the monastery between Rome and Naples which had been the scene of a famous battle during the Second World War. On 29 August they drove to Cassino, where Bruno offered to get Paul from his hiding place. His price, however, had gone up to $3000. It would appear that Bruno took the money and vanished.

At the beginning of September Gail and della Ratta went to England to try to convince Getty to come up with a realistic ransom. They were not allowed to visit Sutton Place and Chace acted as an on-the-spot intermediary between Getty and his former daughter-in-law. Gail felt that Chace was feeding the old man's paranoia. 'Big Paul didn't want to talk to me . . . Chace would talk to him and then meet me.' During this period Gail was kept under surveillance by a private detective, unbeknownst to her. While she was in England her attempt to convince the elder Getty to pay the ransom received a major setback. Chace's fears that the kidnapping might be a hoax were buttressed by Benito Mario Andolfo, formerly Paul Jr's chauffeur and then custodian of Getty's Italian villa of Palo. Andolfo claimed to have seen Paul III in the Piazza Farnese on 15 September. He told the police that Paul III had arrived with another young man and spoke to a third person 'who was pretending to write in a notebook while seated on a garden wall'. Andolfo insisted that Paul III had taken off his sunglasses and 'stared long at him'. Andolfo then asked the young man sitting on the wall to telephone him that evening – which never happened.

Andolfo did not get away with this story because a plainclothed member of the Rome mobile police unit, alerted beforehand by Andolfo, was secretly watching. Indeed, the third person sitting on the wall was himself a police officer in plain clothes, and the police reported that 'the young man indicated by Andolfo was not Getty'.

This bizarre episode was one of the key events of the entire kidnapping. Andolfo was a paid employee of Getty and a personal friend of Paul Jr. He was not considered an intelligent man, but was extremely faithful to his friend and mentor, Paul Jr, with whom he kept in close touch by telephone all this time. Gail always thought that Andolfo was trying to become a hero like Chace and become more valued by his employer. She continues to wonder to this day if the hoax was really an attempt to make the kidnapping appear to be fake – in other words, that Andolfo may have claimed to have seen Paul III as a way of telling the young man if he were only in hiding, and had not really been kidnapped, that his attempt to extort money was bound to fail. The real mystery here is why the police, knowing that Paul III was not sighted at the time, did not immediately call Sutton Place and communicate the truth of the episode, rather than let a false impression delay the efforts to obtain Paul III's freedom. However, even if Getty had known that Andolfo was lying, there is no certainty that he would have paid up more quickly. 'I was angry at Big Paul's fear. He thought the kidnappers were going to come for him. I wish he could have sat down with me,' says Gail now.

Andolfo persisted so adamantly in his false testimony that he was subsequently charged with 'crimes of false witness and slander'. Only after Paul III was freed and denied Andolfo's testimony did he retract, saying that he had erred in good faith. Gail still asks, 'Why did he do that? Did he do it out of loyalty to Paul Jr?' Andolfo has promised to tell her the truth some day.

After Andolfo claimed to have sighted Paul III, Chace left Italy for a few days and was replaced by Bill Newsom. When Newsom landed he was greeted by the Italian press as the 'Godfather of the Golden Hippie'. Newsom's role was to keep contact with the kidnappers in Getty's name and 'ask them for a meeting to reach an agreement', according to the Lagonegro magistrate's report. 'The result was that the kidnappers agreed to meet' with Gail and Newsom on the Rome–Taormina highway. But, Gail claims, Chace advised her not to go. 'I didn't go,' she says, 'and they punished Paul. I regret that I didn't go.' Gail did not know that on 8 October *Il Messagero* had received a letter threatening to send Paul III's ear if she 'did not communicate her willingness to pay the ransom demanded.'

At this point Ed Daly, chairman of World Airways, an American airline, and a friend of Gail's father, flew to Rome to help. Daly arranged a meeting with Archbishop Paul Casimir Marcinkus, the Chicago priest who managed the Vatican Bank and later became controversial for his association with the convicted banker, Michele Sindona, and Roberto Calvi, who was found hanging from scaffolding under Blackfriars Bridge in London. Marcinkus suggested to Gail that she contact the Mafia for help in retrieving Paul III – another recommendation that Chace rejected, for obvious reasons.

Instead, Chace chose to contact the American Embassy's shrewd Federal Bureau of Investigation agent, Thomas Jack Biamonte, for advice on how to proceed. Because the kidnappers expressed irritation with Chace, who spoke no Italian, Biamonte, of Calabrian extraction and a graduate of Brooklyn Law School, took over the negotiations by telephone, all the while pretending to be an American lawyer.

Not only the Gettys but the whole world was shocked when, on 10 November, a package arrived at the offices of *Il Messagero*. It contained a human ear and a lock of reddish hair, which Gail immediately identified as Paul's. It also contained a note demanding 1700 million lira ($3.2 million) and warning that if it was not paid 'the boy's other ear and his body would arrive all cut up in little pieces'. To make sure that there was no misunderstanding, another Rome daily, *Il Tempo*, was invited to send two reporters to a location on the Rome–Naples highway. There they found Polaroid snapshots of J. Paul III standing in front of a cave. He was without his right ear.

The prosecutor's subsequent official report makes it brutally clear that 'not even this mutilation done to the boy' induced the older Getty to pay without delay the requested sum. Instead, an offer of 600 million lira ($1 million) was made by representatives of the family if Paul III was released simultaneously with the payment. However, the kidnappers – who had the upper hand – insisted on 1700 million lira.

At this point, according to Gail, Chace came to see her and told her that Paul Jr was willing to put up $1 million if she would give up custody of the children. Paul Jr denied then that he ever told Chace to make this offer, and Gail believes that it is possible that he did not have anything to do with it. At the time, however, crazy with fear and desperate to save her son, she claims she agreed to give up the children. 'I made the arrangement to put them on the plane,' she says, 'and then it didn't happen.'

Gail made an announcement in the form of an open letter which was meant to be a message for the kidnappers. 'There are hearts like those of Paul's grandfather and father, unhappy because they are petrified and not warmed by love. The grandfather has remained fixed in his loveless old man's arid solitude, but the father has finally accepted to pay a ransom of $1 million under the conditions that you impose.' But the kidnappers would not settle for only $1 million. More would be needed, and J. Paul Getty was the only man who could come up with it. Behind the scenes, Gail's father, Judge George Harris, spent a lot of time on the phone urging Getty to pay. Gail also made a public appeal to President Richard Nixon, hoping, she says, 'to embarrass Big Paul into it'.

In the end Getty gave in. No matter how wild his grandson was in his personal life, no matter how radical his politics, he could not let the barbarians cut more parts from the boy's body. Nor could he bear the adverse publicity. Of the $3.2 million demanded by the kidnappers, he would contribute $2.2 million and lend the remaining $1 million to his son, to be repaid out of his share of the income from the Sarah Getty Trust.

Chace was chosen as the emissary, but he was worried that the kidnappers would pull a gun on him, take the money and shoot him. He wanted someone else to accompany him, a sensible, armed escort. Biamonte found an experienced man and, on the evening of 6 December, Chace withdrew 1700 million lira from the Banca Commerciale Italiana in denominations of 10,000, 50,000 and 100,000 lira notes. Every single note was microfilmed. The next day he drove south to make contact with the kidnappers, but it grew dark and foggy and the highway had been closed because of snow. Chace returned to Rome. The negotiator for the kidnappers, nicknamed 'Cinquante', was furious, and the operation was rescheduled for 12 December.

Chace left Rome at 8 A.M. on 12 December and drove at a constant speed of 70 kilometres an hour along the highway that stretched from Rome, through Naples, Salerno and Lagonegro Nord. Police agents in disguise tailed Chace and his escort to their destination on the highway to Calabria and were able to obtain the registration number of the kidnappers' car, a Mini Minor, which subsequently led to their identification. Chace was heading towards Reggio Calabria when suddenly he heard stones hitting the car, followed by a pistol shot. He stopped the car and saw a man wearing a ski mask and holding a pistol. Chace took the money, which was packed in three bags, from the car and left it beside the road.

The next morning 'Cinquante' told Gail that Paul III would be left on the highway leading to Calabria. While Gail was driving south, however, Paul was found in 'precarious health' by the local police in Lagonegro, on the highway just south of Naples. The sixteen-year-old boy told his rescuers: 'I am Paul Getty. Give me a cigarette. Look, they cut off my ear.' It was 15 December, the day on which J. Paul Getty had been born eighty-one years earlier. With typical self-delusion the old man said it 'was the finest and most wonderful birthday present of my life.'

When Paul III left the police station he was more of a celebrity than his grandfather. Everywhere he went in Rome he was escorted by the police. Girls screamed at him as if he was a Beatle or Mick Jagger.

Only the Italian detectives came out of the affair proud of their accomplishment. They moved in on the tiny desolate hamlet of Cicala in the Calabrian mountains to arrest Antonio Mancuso, a thirty-five-year-old carpenter, whose car had been seen tailing Chace on his drive south to ransom Paul III. The man accused of driving the Mini Minor on the stretch from Naples to Salerno was Saverio Mammoliti, an escaped convict and a well-known drug dealer. He vanished before the police could arrest him. In Rome Domenico Barbino, a twenty-six-year-old hospital orderly, 'the person . . . seen dragging the bags of money' Chace had left on the roadside, was arrested. Then some lira notes used for the ransom were found hidden in a disused stove in the cellar of the home of Giuseppe la Manna, a drug dealer.

For a long time Paul III feared that his life might be at stake if he testified against the men accused of his kidnapping, but he finally made a statement identifying Severio Mammoliti as 'one of the principal executors of his kidnapping'. He also identified Mammoliti as 'the one who, at the end of the long trip which began right after his seizure, had pulled him out feet first from the car and thrown him on the ground.' The prosecutor's report states that 'due to his horizontal position, Getty could have well seen Mammoliti's face, even underneath the blindfold over his eyes.' It concluded: 'There is no doubt that the hypothesis that Getty faked the kidnapping is false.' All the same, after a lapse of twelve years, the old canard that the sixteen-year-old boy arranged his own disappearance in order to extort money from his grandfather is still repeated. There was never the slightest substantiation for any of the vicious gossip and speculation about him or about his mother. Biamonte says that 'if she had been part of the conspiracy she would have been arrested on the spot.' The Italian prosecutor's report was full of praise for Chace, whom it described as 'a precious collaborator, a man of uncommon courage'.

Gail and Paul III had to return to Italy and testify at the trial of the kidnappers who were caught. They turned out to be part of a drug-dealing ring. In their defence they attacked Paul III's character but never accused him of plotting his own kidnapping. However, some of Getty's aides still believe that Paul himself staged the kidnapping, but was then sold to a professional gang who cut off his ear to get the money.

Getty himself seems finally to have accepted that the kidnapping was genuine. He invited Paul III to Sutton Place and personally paid for his later education in California. His personal reaction to the kidnapping was, predictably, an intensification of the fears he had always had for his own safety. He was obsessed with methods of protection, and had had the windows of his Italian villa barred. At Sutton Place there was always a guard outside his bedroom door all night, along with one of the Alsatians. Now his bedroom door was taken off its hinges and a steel sheet embedded in it. An armed bodyguard followed him at a discreet distance when he took his afternoon walk in the grounds. The Surrey estate was ready for the Mafia.

'Paul had problems, no question,' says Gail, 'and he went from being very well behaved, very well groomed, into a total rebel. Of course, his grandfather was absolutely horrified.' The kidnapping, Gail feels, happened because 'Paul got in trouble because of drugs, therefore maybe he bought drugs and didn't pay for them.' Gail feels that none of Paul III's friends 'fingered him; there's no question of that'. One of the habitués of the night clubs that Paul frequented was later arrested as a big-time coke dealer with a fancy home and a large motorboat.

Only $17,000 of the $3.2 million ransom was ever recovered. Never before had Getty been taken for so much money.

18
The Art Lover

An individual without a love for the arts is not completely civilized.

J. Paul Getty

For a man who accumulated oil shares, women and titled acquaintances, collecting art may be said to have been in character. For Getty it was an on-again, off-again addiction that lasted forty-five years. Primarily, it was a way to satisfy his ambition for social recognition as a man of culture. Secondly, the collection was meant to be a way to purchase a benign posterity. If money did not procure an equal stature with the figures of world aristocracy, he could gain their acceptance by collecting the art they respected. 'One way to feel less inferior was to have the art they admired in your house,' says Frederico Zeri who was a close adviser on the collection and until recently a trustee of the Getty Museum.

Another way to feel less inferior was to appear grandiose. Martin Zimet, owner of French & Company, the New York antiques dealers, recognized that Getty saw himself as a king. 'In his mind Getty came close to being king of the world. He only wanted to buy regal works of art.' Anxious to encourage him in this view of himself, Zimet drew up a master list of all the royal pieces of eighteenth-century French furniture that were available from private collections. A desk used by Louis XIV for official business was an acquisition that helped give Getty the semblance of a royal pedigree. Size was important. 'He liked to associate with large objects rather than small ones,' says Burton Fredericksen, who was curator of the collection in the 1960s. 'He preferred Rubens and Titian nudes to small virgins; marbles and bronzes to vases and terra-cottas; huge commodes and tables to side tables and porcelains. He was concerned about size, a man who thought big. He was getting more for his money if he had a big picture.'

At times he would say that art was his pleasure, while work was his duty. 'If I concentrate solely on work,' he explained, 'I not only narrow my perspectives and reduce my effectiveness, but I derive very little enjoyment from living. Ergo, the only sensible solution is to follow a course which permits me to simultaneously work and live to the utmost to enjoy a rounded existence.' This explanation sounds dutiful, but by all accounts Getty certainly got fun out of the quest for art and collecting information about it. His curators and art dealers always found that he would break off from Getty Oil business to talk about the museum or a picture.

Art put Getty in touch with a group of fascinating and colourful characters and created a bond between him and a number of highly talented men and women in the art world. John Brealey, one of the world's foremost picture restorers, calls him 'the most eccentric man I ever knew' and says he had 'more fun from knowing this kind of mad uncle' than with all his other prominent clients gathered together. And they include the Rothschilds and most of the famous ducal families of Europe. Frederico Zeri, who had a flair for hyperbole, called him 'the greatest and most intriguing genius I ever met'.

Ethel LeVane wrote of him: 'He was not . . . merely an individual with acquisitive habits and enough money to indulge his whims. [He was] sincerely appreciative of the arts.' She put the words in the mouth of Mueller, the fat Dutch art expert who was in reality her own alter ego in the book she wrote with Getty, *Collector's Choice*.

It must be recognized, however, that Getty's relationship with art was no less contradictory and inconsistent than his relationship with the oil industry, his women friends or his sons. While he was genuinely excited about his purchases, he always emphasized that 'fine art is the finest investment' and would evaluate the price of each object to determine if he was buying it below its immediate resale value. To LeVane he explained his preference for tapestry, carpets and furniture over paintings as follows: 'Pictures have become too fashionable. People will pay a hundred thousand dollars for a second-rate painting by a second-rate master, and believe they're getting good value . . . In classical statuary, French furniture, tapestries, and carpets I consider I have . . . masterpieces. Yet, if paintings were my preference, I could never hope to compete with the Louvre, the Prado or fifty other fine collections.' His unwillingness to compete with the great museums was a mistake because Getty could have accumulated a wonderful collection of old-master paintings in the 1950s and 1960s and still have seen them appreciate in value more than his furniture or antiquities. Sadly, he was appalled by the then record price of $2.3 million paid by the Metropolitan Museum of Art for Rembrandt's *Aristotle Contemplating the Bust of Homer*.

Bernard Berenson, the elder statesman of the art world, chided Getty in the 1950s for not paying up for old masters in order to acquire a great collection. 'Thus far you have been getting a few pictures of the kind the Kress Foundation sends to every town where it has a five cents store,' Berenson wrote in a disparaging fashion. He advised Getty that he would have to spend at least $300,000 to get a fine picture and recommended going to dealers like Thomas Agnew and Colnaghi in London, and Knoedler, Duveen and Wildenstein in New York. Nevertheless, Getty antagonized the owners of Wildenstein when he doubted the pedigree of a Botticelli and a Titian that had been judged masterpieces by Berenson.

According to Zeri, when Berenson died, Nicky Mariano, Berenson's long-time associate, wrote to Getty requesting a donation to endow Berenson's villa as a fine arts study centre for Harvard University. Getty sent a cheque for $1000, which so infuriated Mariano that she returned it.

He seemed to be more enthusiastic about beautiful carpets than any other art form. Gillian Wilson, his decorative arts curator, remembers him 'down on his hands and knees inspecting an oriental carpet at Sutton Place, crawling along describing the knots and dyes to me. He was obviously fascinated by it, loved it, and wanted to show and talk about it. I call that a passion.'

Another inconsistency was his taste for both the highly elaborate French furniture of the eighteenth century and the simplicity of classical statues. He would collect, say, a delicate charming boudoir scene painted by François Boucher 'reflecting the feminine influence of Pompadour and Du Barry', as well as the forceful, massive Lansdowne Hercules. He wrote about the 'joys of collecting', yet measured each purchase as to whether it was a good investment or not. He would study the X-rays of a painting to determine its origin, but reject it if the cost per square inch seemed too high. He prided himself, with justification, on his scholarship, yet was afraid to purchase many paintings until after a prominent gallery had bought them first at auction. Sir Geoffrey Agnew, the London dealer, believes 'he wanted to be thought of as a great collector without taking responsibility for it', while Peter Wilson, the late chairman of Sotheby's, said he 'didn't have the flair for collecting paintings'.

Getty's appreciation of art probably began when he was an impressionable young man in Florence's Uffizi Gallery standing in front of Titian's magnificent *Urbino Venus*, a naked woman lying in the most inviting fashion on a couch. She is bathed in a sensual warm golden hue and, despite holding her hands in a modest pose, seems to be beckoning the viewer to join her. No doubt the naked female body in art was a powerful attraction – Getty used to touch the bodies of women depicted in Roman and Greek statues with the full palm of his hand. Perhaps he fantasized about possessing these women, for he used to point at Joshua Reynolds's portrait of Joanna Leigh, an especially lovely woman, and tell his guests, 'There's one that got away.' He was probably led astray by the curvy body of a fake Venus he bought in 1939. Jiri Frel, until recently curator of antiquities at the Getty Museum, says, 'The Venus was probably made in the early twentieth century and is entirely too suggestive for a piece of classical statuary.' An article Getty wrote for *Playboy* magazine showed him rebelling against the puritanical values of his parents:

There are, I suppose, several principal reasons for the indifference – if not open hostility – of the majority of American men toward all things that come under artistic or cultural headings. Some of the roots can be found in our Puritan heritage. Early American Puritans, hewing to their stern, super-Calvinist doctrines, equated art with depravity, branded most music as carnal and licentious, shunned literature other than religious tracts or theological discourses and condemned virtually all cultural pursuits as being frivolous and sinful. In the Puritan view, that which was not starkly simple and coldly functional was, propter hoc, debauched and degenerate.

In the 1930s Getty learned about art as luxury when he lived among Mrs Frederick Guest's antique furniture and carpets in the penthouse apartment he rented from her on Sutton Place in New York. Prominent American families surrounded themselves with the most elegant artefacts of European culture to demonstrate they were not wealthy barbarians, and Getty began 'to appreciate that a table, a chair, a cabinet or a commode could be as much a fine piece of art as a painting or a sculpture.' Methodically, he had himself tutored in the appreciation of works of art, specializing in French furniture of the eighteenth century, the 'golden age of furniture', in which 'France reigned supreme'. It helped 'to bring a bygone era to life' and invigorated him, 'adding breadth and depth to my whole existence'. In his book *The Golden Age*, his personal prescription for living a full life at seventy-five years of age, he describes 'the drama, adventure and thrills one experiences when one succeeds in acquiring a superb chair, divan, desk or other piece of furniture that is not only an outstanding work of art, but was once owned or used by a King of France, a Marie Antoinette, or a Pompadour. Believe me, history and its great figures do come to life.'

By the late 1930s Getty realized that the objects of past civilizations could be acquired for a fraction of their value, due to the impact of the economic depression and general fear about another world war. At some of that period's great sales he picked up the beginnings of his fine collection at rock-bottom bargain prices. His brilliant and opportunistic buying at the Mortimer Schiff sale in 1938 created the nucleus of his French furniture collection. For ever after he would describe his collection as 'Before Schiff, Schiff and After Schiff'.

In the early 1950s, when most newly wealthy American collectors preferred French Impressionist paintings, Getty moved in rapid succession to buy three sculptures from the Earl of Elgin's antiquities collection at Broom Hall.

He bought only a handful of mediocre Impressionist

paintings in the 1950s before their prices took off into the stratosphere. Furniture, carpets and antiquities were still far cheaper and they remained extraordinary bargains until much later. He was buying art that the Greek shipping tycoons and newly rich American millionaires did not want. 'He could purchase a fine Louis XV piece for $30,000 when a Cézanne was going for $600,000,' says Claude Sere, a Paris antiques dealer. It was during the 1950s that he acquired the so-called husband-and-wife desk by the eighteenth-century master Bernard van Riesenberg, an exquisite piece. 'Without devoting large sums of money and publicity to his collecting, he has succeeded in assembling . . . some of the most precious specimens of the work of the Parisian cabinet makers of the eighteenth century,' Pierre Verlet, then keeper of the Louvre's department of furniture, wrote years later, adding that Getty 'comes close to being the ideal art-lover, for flair grows through experience, planned or otherwise, and out of hard work, whether it be conscious or unconscious. Perhaps geniuses do exist, but one is more inclined to trust in the energy and intelligence of certain men who are, to be sure, gifted with a natural finesse, but who are above all devoted to the acquisition of knowledge . . . allowing them to hunt down the noblest and most unusual quarry with confidence.'

Unfortunately, Verlet's esteem of Getty as a collector was not followed up to gain the American oilman as a benefactor to the Louvre. Getty became enraged at Verlet for not displaying a bejewelled snuff box once belonging to Louis XIV, the Sun King, which he had given to the Louvre. Paul Louis Weiller believed the gift to have been 'a beautiful gesture. Paul was terribly angry and resentful that a snuff box donated by a Rothschild had been given prominence. He would have been very generous and ready to donate anything else.'

For Getty, the giving away of art was closely related to the constant nagging problems of his taxes. From 1948 on, Getty used the donation of art objects as a way to shelter his income. Art became more important than the oil depletion allowance and other tax loopholes utilized by men whose livelihood was exploring for oil. The gifts Getty gave to the Los Angeles County Museum between 1948 and 1953, which included the Coronation carpet, a Persian rug of the Safavid dynasty, Rembrandt's portrait of Marten Looten and a Beauvais tapestry, were appraised at a value of $700,000 and deducted from his taxes during those years. By 1974, Getty's contributions to the J. Paul Getty Museum totalled $9,311,592 or roughly half his gross income. Then in 1975 he gave $14,733,654 – again almost half his gross income of $29,467,308. In this way, Getty substantially limited the tax bite from Uncle Sam. In 1975, for example, he paid only about $4.2 million or a 14 per cent tax rate. Not bad for the richest man in the world.

Getty's taste was related directly to the fascination he felt for historical periods that provided him with the dream of a better, happier life. Typically, he admired the superficial elegance and order of the French court society without considering its corruption and decadence. And he often seemed to be on the defensive about the quality of his collection, even before the attacks on it began. In 1965 he wrote in the foreword to *The Joys of Collecting*: 'I have tried to keep the level as high as possible for it seems to me that, human nature being what it is, a collection is often known by the worst thing in it rather than the best thing in it' – a very questionable statement. But he failed to put together a consistently sublime collection, like Frick or Mellon. Even his paintings expert, Julius S. Held, a professor of art history at Barnard College, wrote in *The Joys of Collecting*: 'It would be less than ingenuous to say that this collection forms a harmonious whole. There is a certain unevenness both in the distribution of schools and in the level of artistic performance.'

Getty was limited by his obsession that works of art should be bought at bargain prices, although he was occasionally inconsistent even on this point. He paid some $400,000 for *Diana and Her Nymphs Departing for the Chase* by Rubens, a huge canvas with a number of typical, bare-breasted Rubens women, one of whom is being attacked by a satyr. But this was an exception.

'I don't think he looked at things and saw they were beautiful. He was interested in technique, in the epoch in which they were made. Most importantly, was he going to get it cheap?' says the Duke of Bedford. The stories about his mania for cheapness all have a similar ring. He sent a piece of sculpture back to an antiques dealer because its cost was too high in relation to its size. He paid the smallest fees possible to art experts who authenticated a work. 'The act of being mean gave him great pleasure,' says Sir Francis Watson, one of his advisers on the decorative arts. 'Getty's collecting was a para-artistic activity. He acted up to the image of very rich men who are collectors of money, and works of art come second.' Only when his income began multiplying by geometric progression with the price of oil in the early 1970s, and under pressure from his fellow collectors like Baron Thyssen and Paul Louis Weiller, did he at last start buying some great paintings. Perhaps the possibility of tax relief gave him a shot of adrenaline when it came to buying pictures, because donating expensive pictures sheltered his accelerating income. Sir Geoffrey Agnew remembers John Brealey coming to his gallery one day with Getty and whispering to him, 'Show him a real work of art, because he's got to spend money for tax reasons.'

Still, Getty complained to his friend Thyssen around this
time that he had come to dinner in a pair of white
sneakers because 'I have given everything to my family!'
Thyssen, who was always amused by his friend's pose of
poverty, murmured, 'Poor Paul, I'm sorry for you.'

Despite his penny pinching, he managed to acquire
some great paintings. In addition to *Diana and Her
Nymphs* by Rubens, which hung in the Great Hall at
Sutton Place, he bought Rembrandt's portrait of St
Bartholomew at a cost of $532,000, and Van Dyck's
portrait of Agostino Pallavicini. But his collection of
paintings, taken as a whole, was considered in reputable
art circles as a 'study collection', which is a polite putdown
and was applied to several hundred paintings he had
bought at bargain prices. Sir Francis Watson thought that
'all but ten paintings should be sent back to their original
owners for free.'

Evidently Getty took it to heart when, one night at
Thyssen's villa, his host told him, 'Stop buying this
rubbish and buy a good picture for once.' Getty turned to
Rudolf Heineman, one of Europe's foremost dealers, and
said, 'I want you to buy me a collection like Heinie's.' Of
course, such a feat was impossible; Getty would not pay
up on such a scale. But Getty was seriously in the market,
at least temporarily, and the high point of his acquisitions
came at Christie's auction house in London on 25 June
1971, when he spent $6 million on four paintings. The
bidding was done on his behalf by the colourful, hard-
selling dealer Martin Zimet, who no doubt prompted
Getty to let him act for him.

Never before had Getty spent $6 million at a single
sale. The four paintings he acquired were Titian's *The
Death of Actaeon*, Van Dyck's *Four Heads of a Negro*
and a pair of Boucher's, *La Fontaine d'Amour* and *la
Pipe aux Oiseaux*. Fredericksen, who accompanied Zimet

to the auction, thought it 'was completely out of character for Getty'. He claimed that every year he donated at least 90 per cent of his taxable income to charity. This was accomplished by giving away works of art, many revalued at higher prices than he had paid. In theory he could deduct the whole of the $6 million he had spent to avoid income tax on the $11,700,000 cash dividends he personally received in 1971 from Getty Oil Company.

Unfortunately for Getty, his purchase of the Titian for over $4 million, up to that time the second highest price ever paid for a single painting, led to an uproar. Thyssen's advice to 'buy a good picture for once' had backfired. The Titian belonged to the Earl of Harewood, Queen Elizabeth's cousin, and its impending departure from Great Britain set off a great storm of protest. Getty offered to leave it hanging in the National Gallery, but the British decided to raise the $4 million to keep it at home. He complained that he lost 6 per cent interest on his $4 million for the year it took to raise the funds to buy it for the nation. 'In effect I was obliged to lend it to the National Gallery interest free for a year,' he said. 'The British did wonders collecting all that money, but at 6 per cent my loss was more than £100,000 (about $240,000).' Sir Francis Watson says, 'Getty acted up to the image of himself. He was an eccentric not a patron of the arts. The richest man in the world doesn't complain about 6 per cent interest.'

Although he wanted to be considered a rich American with a taste for culture, a model of taste, who was not devoted 'exclusively to the worship and pursuit of the almighty dollar', few, if any, observers ever saw him exhibit, physically or emotionally, any great feeling about a work of art. Stephen Garrett, the British architect, who became deputy director of the J. Paul Getty Museum, says, 'There was no *joie de vivre*. I never saw him stand in front of an art object and enthuse about it,' and although Zeri describes how 'tears came to his eyes from a simple brushstroke', it is easier to believe dealers like Rudolph Forrer of the London company Spink & Son, who said Getty was 'the most difficult of our customers . . . He never expresses his feelings, never seems carried away with the beauty of an object.' Perhaps he never showed any excitement because he did not want to be exploited. It was well known that the art world was rampant with sharp and unscrupulous dealers, but in old age he became inured to the dealers who streamed to Sutton Place showing their wares. When they told him, 'It's a good buy, Mr Getty, it's a good buy,' he would sometimes reply with a frail wave of his hand, 'Goodbye, goodbye.' One doubts his sincerity when, in the foreword to *The Joys of Collecting*, he asserts that 'the true worth of a collection cannot – and should not – be measured solely in terms of its monetary value.' Sir Geoffrey Agnew says, 'I find it difficult to say how much he liked the things.' If there was passion, it was expressed in his incredible drive to authenticate every piece of work. He steeped himself in its history and its physical composition, hunting down the details in the museums or the great dealers of London, like Thomas Agnew & Sons, where he might spend an entire afternoon poring over his research. He liked to collect every scrap of information about a work of art from the art historians and dealers who trooped down to Sutton Place, where he would keep them up till all hours of the night bombarding them with questions. Only when there was nothing more to squeeze out would he let the expert go to sleep. 'Then, he knew as much or more than you did, and he was happy,' says John Brealey, 'because then he had the advantage. It gave him great pleasure. He knew as much about technique as anybody. But he didnt really love them.'

Although he was devoted to Getty, Brealey felt his client had no understanding of art at all. 'If you put him with a group in front of the Acropolis, everyone else would be on their knees. He would ask how the foundations were made.' Brealey was flabbergasted and amazed by the way Getty's mind worked. 'He would ask one simple-minded question after another. They had nothing to do with the grandeur of the picture or the importance of the work.' To Brealey, Getty was 'like a piece of blotting paper. He wanted information. He literally wanted to suck knowledge out of the experts,' says Brealey. Yet his naïveté was 'unparalleled. It was as if he had never grown up. I failed to understand where his abilities lay. I was being entertained by this extraordinary puzzle. I felt I was in pursuit of the truth but it evaded me.' In the end, Brealey simply concluded that Getty was 'a country bumpkin, a hick, a yokel who had become extraordinarily rich.'

It was Brealey, one of the world's leading picture restorers, who told Getty that *The Madonna of the Loretto*, which he had picked up for $200 in 1938, could not have been painted by Raphael, as Getty wanted desperately to prove. He had an obsession about this painting, and the story is remarkable, particularly as it involves one of the last major love affairs in his life, and one of the strongest.

Brealey, then at work on Mantegna's *Triumph* in the Royal Collection, told Getty that the blue used in *The Madonna of the Loretto* was an inexpensive substance made from azurite, rather than the more costly blue colour made from lapis lazuli. This really worried Getty. How could Raphael have used a cheap blue in Getty's picture? It was not the colour itself that mattered. It was the fact that it was cheap. How could the richest man in the world have a picture with a cheap pigment? Getty set about researching the cost of pigment during the first quarter of the sixteenth century and was able to tell Brealey with glee some months later that azurite cost more in 1508, the year the picture was painted, than lapis lazuli.

Not that this proved the matter one way or another. It was known that the original Raphael *Madonna* had hung in the Roman church of Santa Maria del Populo until 1591 when Cardinal Sfondrati decided to expropriate it along with another Raphael masterpiece, his portrait of Pope Julius II. In 1608 Cardinal Sfondrati sold his collection to Cardinal Scipione Borghese and the picture vanished from public record, except for the fact that when Cardinal Borghese's collection was inventoried the painting was given the number 133.

It was the hunt, the pursuit of information, that Getty obviously loved. If he could prove that his *Madonna* was the authentic missing Raphael masterpiece he would be hailed as a genius in the world of art, and would incidentally have made an extraordinary pile of money on one picture. Unfortunately, Getty's *Madonna* had deteriorated physically to such an extent that no one in the art authentication business could tell for certain whether it was *the* Raphael, or even *a* Raphael. Zeri said, 'How can you judge a picture that is destroyed?' However, he added that the infrared photographs of the painting disclose 'a drawing that shows the influence of Leonardo da Vinci'. And Leonardo influenced Raphael. Undoubtedly the underdrawing of the painting was the work of 'a great artist', says Brealey. 'It was closer to Leonardo in style than Raphael. Whoever made it was a great master. It electrified Getty and had a mesmeric effect on him so that he made it his life's work.'

By the mid-1960s the painting of the tranquil virgin lifting the veil from the awakening infant Jesus, while

Joseph stands solemnly in the background, had been cleaned, X-rayed and received the stamp of approval from art historian Alfred Scharf, Cecil Gould, deputy director of London's National Gallery, and a host of others as the likely original *Madonna*. For many years the *Madonna* sat on an easel in Getty's bedroom opposite a rather mediocre Renoir, one of the few Impressionist works of art he owned. Then he offered the picture on loan, along with a portrait by Paolo Veronese, to the National gallery, where it hung for many years. Cecil Gould says, 'When the loan of both was accepted, it flattered his vanity that something he had bought so cheaply . . . should now be on display at the National Gallery and labelled as Raphael. He admitted that the picture had come to have such a hold over him that he wanted it out of the house.'

From 1965 until 1973, Getty reverentially visited his Raphael at the gallery, bringing his friends to gaze at it. He would spend hours there staring at the painting which he had bought for $200 and which was now, he boasted, insured for $2 million. He spent as much time at the British Library, reading every conceivable work about Raphael, the period of history and the painting. It added international lustre to his reputation when the painting was shipped to the United States for exhibition in the Metropolitan Museum in New York and the National Gallery in Washington.

Did he really believe he owned the Raphael *Madonna*? Ethel LeVane quotes him in *Collector's Choice* as saying, 'Mine is undoubtedly only a copy. There's not one chance in a million that it's the priceless Raphael.' Nevertheless, he never stopped working in his attempt to prove that, of the thirty-four *Madonnas of the Loretto* that existed in the world, he and he alone owned the authentic Raphael version. He personally went over every pore and crack in the painting looking for the number 133, without success. Curiously, he never seems to have investigated the other thirty-three versions, including one in Chantilly in France.

And now a mysterious lady from Czechoslovakia enters the story. Her name was Anna Hladka; she was forty-two years old, married, with two young sons; and she was a journalist working for the Czech News Agency. One day in 1965 she was praying in the Church of Little Jesus of Prague when she heard a voice say. 'You will meet a man called Goethe, who lives in a place called Alsho.' On the very day that she heard the voice, Associated Press ran a story about Getty's purchase of Palo, the Italian villa on the coast near Rome. She instantly related the name Getty to Goethe (a connection the oil billionaire would have loved). Palo, she came to learn later, was near Alysium, the Roman port where Caesar landed after his African campaigns and, to her, that seemed close enough to 'Alsho'.

She became fixated on Getty and Palo, researched the villa's history, and finally, in the autumn of 1966, while her husband Ladislav was ill with cancer, she poured out her heart, together with her researches, in a letter to Getty at Sutton Place. His social secretary at the time, Emma Neville, the god-daughter of Lord Louis Mountbatten, automatically put the letter in the 'Nuts' file. Nevertheless, Getty answered Hladka with a brief but personal note, thanking her and opening the door to further communications. 'I do hope that we can meet some day and I would enjoy a letter from you,' he wrote and, incredibly, signed it 'Paul'. He enclosed a photograph of himself standing near the cropped yew trees with the impressive bulk of Sutton Place looming behind him. The lord of the manor seemed to be offering a haven from socialism and an unhappy life, and indeed Hladka, as she said later, wanted 'to live in freedom once more'. She knew England, having worked at the Czech Embassy in London in 1947.

With her next letter Anna sent a small photograph of herself standing in a doorway, with the wind blowing through her dark hair. She was not beautiful, but sensual and with a half smile that intrigued Getty. It was the beginning of a regular correspondence, on her side full of emotional messages, on Getty's polite, devoid of feeling, but definitely containing invitations to visit him. In 1967 Anna arranged a job as a cook in a summer camp in Wales. She bought a round-trip ticket from Prague to Guildford, Getty's nearest railway station, and Getty put her up at Sutton Place.

He must have found the small, poorly dressed woman attractive. One afternoon they were sitting alone on the settee in his study. Suddenly he 'put his hand on my knee', she told the author. She at first 'thought to slap this old man in the face', but caught hold of herself and asked him with a blush, 'What does an English or American lady do when you put your hand there?' Getty laughed and, in his deep, slow voice, answered, 'She would be pleased to have a new customer.' Anna was even more shocked by his earthy response. She felt strongly, 'I shouldn't trust this man.' There was no further intimacy until the day she was to leave. Getty said, 'You are like chocolate,' and kissed her. She asked him to help her obtain political asylum in Britain, which would require her getting a divorce and custody of her two sons. When she returned to Prague Getty asked her to investigate the purchase of pipeline equipment from the rather more liberal regime ushered in by Alexander Dubček.

During the summer of 1968 Anna felt herself in great danger as large numbers of Eastern Bloc soldiers massed on the Czech borders. As the world anxiously watched for a Soviet invasion, Getty's notes to Anna sounded like recommendations to flee. On 12 July he hoped 'you enjoy your visit to the sea coast'. There being no sea coast in Czechoslovakia, this may have been an invitation to her to cross the Channel and join him. On 6 August he told her she 'will be invited as a friend, not as a worker', and a week before the Russians marched in, on 14 August, he wrote: 'I would be glad for your visit as soon as it is convenient for you.' After the voice in the church, Anna believed her saviour was 'summoning me'. On 28 August, a week after the Soviet invasion, she took her two sons and made her way secretly to the British Embassy, where officials checked her story with Sutton Place. There the summer guests, who included Mary Teissier, Zeri and Ronald Getty and his family, were incredulous when this strange, dark Czech refugee arrived with her two children.

Anna Hladka became Getty's full-time researcher. He gave her a roaming commission to track down every obscure lead about the *Madonna* and to answer his inquiries about the history of Palo, as well as Sutton Place. She went to study archives in France, Italy, Spain and Austria, hunting for the documentary proof that Getty's was indeed the authentic *Madonna*. She realized early on that 'he made work for me to do, because all the books in the British Museum on the *Loretto* were marked with his notes in the margins. Slips of paper with his name were in the books.'

One day in 1972, while doing research in the British Museum's Department of Prints and Drawings, Anna came upon some disturbing information. On a loose-leaf file that held engravings of Raphael paintings there was one of *The Madonna of the Loretto*, labelled: 'This is the picture by Raphael which is at Santa Maria del Populo ML', the church in Rome where the original *Madonna* had originally hung. To Anna's eyes, the placing of the three figures – Virgin, child and father – were not identical to that of the figures in Getty's painting.

The engraving was dated 1572 and on the back the number 965 had been pencilled in. This referred to the Windsor Castle Raphael Collection inventory, but when she consulted the catalogue she discovered that 965 was entered as another engraving of the *Madonna* by J. T. Richomme, a Frenchman, made in 1812 from a painting by Giulio Romano. It appeared that the 1572 engraving – which Anna thought disproved the authenticity of Getty's painting – had not been entered in the inventory at all.

Getty had by this time bought Anna a small house in Sussex, about 35 miles from Sutton Place, and from there she wrote to him on 1 November 1972, suggesting the possibility of obtaining an engraving by Richomme from Paris to replace the 1572 engraving. After all, the British Museum thought it had only the Richomme engraving. This ploy might have enlivened their relationship, but it is doubtful if that single confusion could have made much difference in establishing the authenticity of Getty's painting. Doubts about it were expressed during Getty's lifetime by art historians like Sir John Pope Hennessy, but it was never cleared up until after Getty's death. In fact, Burton Fredericksen, who had become curator of his paintings, discovered in 1973 the Borghese number 133 on the painting at Chantilly. But he was uncertain about its significance and did not publish his discovery that the Raphael *Madonna* was at Chantilly until after Getty's death in 1976.

Anna continued to do research for Fredericksen on the *Madonna* but after Getty's death she was 'let go' by Sutton Place Properties, her employer. She appealed to a British industrial tribunal but lost and, with a small pension in hand, went to work for the firm which arranged to insure the transportation of Getty's pictures to Malibu.

By the early 1970s Getty was creating the unique museum at Malibu, in southern California, which was to replace the ranch house where his collection was on display. He had bought the 64-acre citrus ranch after the Second World War. His death and the still secret legacy of his entire personal fortune to the museum would, he knew, give rise to several problems. The ranch house was already filled with his collection and space had to be created for all the Sutton Place paintings and furniture. And the steady increase in his personal wealth led him to expand his collection during the period 1969–71. In any case, it was only open two afternoons a week, watched over by Burton Fredericksen. There were rumours that the Internal Revenue service had told Getty's lawyers that they might question the museum's tax-free status unless it became a fully fledged charitable institution, regularly open to the public.

But what sort of building was to become the permanent home of his treasures? He told the trustees unequivocably, 'I refuse to pay for one of those concrete bunker-type structures that are the fad among modern architects – nor some tinted-glass-stainless-steel monstrosity.' No Guggenheim Museum, no Museum of Modern Art for him. He preferred ancient grandeur to modern luxury.

Throughout his life, Getty identified with the Roman emperors. He read about them and from time to time fancied that he was like them. Brealey even felt that Getty looked 'like a Roman emperor in profile'. He identified with the power and the majesty of what they had achieved and he wanted to achieve more if possible.

He determined to create a unique museum which would actually be a rough re-creation of a luxurious villa of Rome's early Imperial period. As a young traveller, he had been fascinated by the Villa dei Papyrii, which was located at the edge of Herculaneum, the town near

Pompeii which had also been destroyed by volcanic ash from Mount Vesuvius. This villa, thought to have belonged to Lucius Calpurnias Piso, Julius Caesar's father-in-law, stretched 800 feet along the Bay of Naples facing the islands of Capri and Ischia, two favourite watering spots for the emperors of ancient times as well as for today's tourists. Piso had been one of the period's greatest collectors. Some ninety pieces of bronze and marble sculpture were recovered from the ruins of his peristyled gardens, including the god Pan having intercourse with a she-goat and a brazier held up by three satyrs with erect penises. The scrolls found in Piso's library may also have fascinated Getty. They included lascivious verses and racy epigrams by the Epicurean philosopher Philodemus, and if Getty was unlikely to observe the Epicurean teaching to avoid the desire for power and glory, at least he could identify with Philodemus's advice to satisfy sexual desire in a casual manner without taking it too seriously.

But it was the building rather than the philosopher that mesmerized Getty. Back in 1955 he had declared that the villa was 'so grand and extensive, and held so many art treasures, that it is difficult for anyone of later generations to imagine such wealth and splendour.' It was a grandiose idea to try to recreate a villa that had been totally destroyed by 60 feet of lava and rock in 79 A.D. and had never been totally excavated. There were no precise floor plans; nor was anyone even sure that the villa had actually belonged to Calpurnias Piso.

Sometimes Getty had a casual way of making a major investment. One Friday night in 1968, at dinner, he simply asked Stephen Garrett, a British architect in charge of renovating his two Italian homes, to look into the possibility of re-creating the Villa dei Papyrii in

Malibu, California, on the same site as his Spanish ranch house.

Garrett was a pleasant, handsome architecture graduate of Cambridge University, who had first been hired to oversee the necessary renovations of Gaiola, the small islet in the Bay of Naples that Getty had purchased from Baron Langheim. As Getty did not want to face the 50-foot boat trip across the Mediterranean to the islet, he instructed Garrett to draw up plans for either a bridge or a tunnel. 'It was unexpected and rather fascinating,' Garrett felt. 'Such was his wealth he could do anything.' Unfortunately, Gaiola had been formed from ancient lava from the eruptions of Vesuvius and was 'eroding like crazy'. Both Palo and the buildings on Gaiola had to be renovated constantly because the salt water had such a deteriorating effect. But Garrett did not mind; he loved the trips to southern Italy to look after Getty's homes.

Some of Getty's ideas he found 'extremely scary'. From 6000 miles away he paid the same painstaking attention to detail as if he was in Malibu himself, supervising every stone. He insisted that the eucalyptus trees around the ranch house should not be moved, and that the museum's design steer clear of them, He had a model of the museum in his study and was sent pictures, even films, of the building as it progressed. True to his nature, he reviewed every single invoice of the museum's construction and, on seeing a bill of $17 for an electric pencil sharpener, he cabled Garrett to desist from such expenditure without his approval.

He wanted his museum to be a unique building feat, saying at the time, 'There is, I believe, no other place in the world where one can go to see such a building in any state except in ruins, as one sees them now in Pompeii. There are replicas and imitations of ancient public buildings, but none of a private structure – so this should provide a unique experience.'

The museum was to be placed at the summit of the Malibu site, where a Corinthian colonnaded loggia opened to the south on the broad vista of the Pacific Ocean. A formal Roman garden enclosed by a colonnaded peristyle led up to the two-storeyed villa, which was meant to have the scale of a domestic dwelling rather than that of a public institution. To duplicate the flavour of a first-century villa, the ceilings are painted in the soft pastels of Rome, the walls are sheathed in marble and onyx shipped from abroad, and the floors are modern copies of ancient mosaics and tile designs. 'I would like every visitor at Malibu to feel as if I had invited him to come and look about and feel at home,' Getty said, imagining that the villa–museum was indeed his home and that he was Calpurnias Piso, or some twentieth-century version.

And yet, astonishingly, although it was his personal creation, Getty never once visited the museum. Nor did he seem any of the objects it housed, only photographs of them, because after purchase they were sent to Malibu and not to Sutton Place. Starting in 1971, most of the best paintings, like *Diana and Her Nymphs* by Rubens, were gradually shipped from Sutton Place to the museum. 'He didn't live with them. He never saw the things,' says Gillian Wilson.

It is a fact, however, that Getty ranked high as a private collector of antiquities and French furniture. His collection of antiquities is considered the third most important in the USA, after the massive collections at the Metropolitan Museum and the Museum of Fine Arts in Boston. His French furniture is the finest private collection in the nation. This was gratifying to a man who thought that 'few human activities provide an individual with a greater sense of personal gratification than the assembling of a collection of art objects that appeal to him and that he feels have a true and lasting beauty.' Claude Sere said, 'He would have been a prince in another age. He was better than the sixteenth- or seventeenth-century princes. Look at what he accumulated.' Brealey, however, found Getty to be 'enigmatic about art as well as every other aspect of his mind'. He felt that Getty's genius lay 'in very close study of the present moment to foresee tomorrow' and considered him 'a tremendously disciplined man. Art was one of the compartments of his discipline and his knowledge.'

Getty's butler, Bullimore, told Garrett he thought the master of Sutton Place was intending to return to California to be at the museum's opening on 16 January 1974, because 'the old man had been up to see his dentist on the National Health, and had arranged for letters of credit.' But he simply could not face crossing the ocean. He was physically afraid and possibly anxious about how the museum would be received by the architecture critics, whom he knew to prefer modern design. He knew that 'the flouting of conventional wisdom and refusal to conform carry with them many risks . . . However, I had calculated the risks and – I say this with an admitted degree of arrogance – I disregarded them. Thus, I was neither shaken nor surprised when some of the early returns showed that certain critics sniffed at the new museum.' One critic said the museum was 'a multimillion dollar piece of unintended folk art'. Others compared it to Disneyland, Forest Lawn and other kitschy, vulgar places that people actually ridiculed. The *New York Times* architecture critic, Paul Goldberg, called it 'pretentious and somewhat sterile'.

This $17 million investment, to house a $200 million art collection, was given a $55 million endowment and secretly promised the bulk of Getty's estate. 'You can't read him in normal human terms,' says Garrett. 'He

spent $17 million on a museum and didn't come to see it.'
He refused to ask for public subsidies in order to be free
of governmental restraints. He claimed his overriding
consideration was to have his 'collection . . . completely
open to the public free of all charges'. How totally out of
character for Getty to foot the bill for ever on behalf of
every visitor to the museum. He had worked out the cost
per visitor according to Garrett, although at first he did
not want to pay for guards. Every dollar contributed to
the museum, of course, was a charitable deduction from
his taxes. So that the art objects at Sutton Place and Palo
would not be subject to foreign taxes, Getty set up Fine
Arts Corporation, a Delaware company, which owned
them before they were given to the museum. He knew
that his 4 million shares of Getty Oil, worth over $500
million in January 1974, would make his museum three
times richer than the Metropolitan Museum of Art in
New York. Nevertheless, he wanted to keep the curators
under his tight puritanical discipline and told them there
would be no more money, thus slowing down acquisitions
after the museum opened in 1974.

Garrett became deputy director of the museum. He
was surprised how wounded Getty was by the criticism.
'Paul seriously considered moving the museum from
here,' he said. 'He considered picking it up from its
foundations and moving it somewhere else like Minneapolis.' There is no doubt that Getty was mortified by the
reaction of the cultural establishment. 'You can criticize
a man's wife, his children, in fact, almost anything about
him,' he had said revealingly years before, in 1957, the
year he became the richest American. 'But, if you criticize
his taste in art, you offend him mortally.'

19

His Prominence

I believe that the able industrial leader who creates wealth and employment is as worthy of historical notice as the politician or the soldier who spends an evergrowing share of the wealth created by individual initiative and courage.

J. Paul Getty

Late in life Getty liked to say that the Gettys were a 'prominent family – almost as prominent as the Kennedy family or the Rockefeller family.' It is true that, by 1974, he had amassed far more wealth than either of them, but he was talking of the honours which he wanted for himself and his family. Unfortunately they had neither the political eminence of the Kennedys nor the great charitable reputation of the Rockefellers. Getty's charitable contributions were negligible relative to his wealth, except for the $55 million museum endowment.

The comparison was also inappropriate because both the Kennedys and the Rockefellers existed as families, meeting together from time to time. By contrast, the Getty clan was a collection of characters who had little or nothing to do with one another. Of the surviving sons, Ronald had given up oil for Hollywood, where he helped produce some not particularly ennobling films, including *Flare-up*, starring Ursula Andress. In 1969 Fini brought suit in Los Angeles Superior Court charging that her son, Ronald, had forced her out of her home in Beverly Hills. He had also upset his brother, George, by using the family name when he set up the Getty Financial Corporation, which owned a chain of Don the Beachcomber restaurants. To block Ronald from using the name Getty

Financial Corporation in any place but California, George ordered Evey, his chief aide, to register the name on Getty Oil's behalf in every other state.

Paul Jr, a sick man, caught in the anguish of the drug culture and troubled by the tragic death of his second wife, Talitha, was fast becoming a total hermit in his London house. Paul had been developing an exquisite taste and broad knowledge in antiquarian book bindings. He was considered to have a very fine collection and offered to donate it to the Getty Museum, but his father refused. The youngest brother, Gordon, who had tried partially to liquidate the Sarah Getty Trust, spent a great deal of time composing music in his San Francisco home, giving concerts and serving on charitable boards. The Nixon administration approached him and offered him an ambassadorship in one of the Low Countries, Holland or Belgium, if he would make a large campaign contribution. However, he did not feel such a huge payment was appropriate and offered a far smaller donation to Nixon. Both Paul Jr and Gordon were among the superwealthy without ever having had to work for it. They were the passive recipients of the wealth created by their father and by 1974 their annual incomes from the trust were approximately $600,000. In a single decade their incomes had jumped more than tenfold.

When it came to prominence, there was only one centre of attention in the Getty family, and that was its eighty-two-year-old head, with his thin, stooped body and wizened face. He spoke slowly in a deep, hollow voice that sometimes had a Midwestern inflexion, perhaps put on. Behind his back his friends loved to mimic the way he spoke. In a television commercial he made for the brokerage firm E. F. Hutton, the old man shuffles to a large wooden door, needing help to lift the metal bolt, and steps outside to face the camera. The birds are tweeting sweetly as he mechanically reads his lines without a scintilla of emotion. 'They have been helpful and their services have been good.' But he was not as feeble as he looked, either in mind or body.

As for charity, he sent a standard letter to every person who wrote to him asking for money. They got an apology instead of a handout. Indeed, the letters were headed 'I apologize', and explained: 'I don't have large sums of ready cash not required for my business.' The letter complains that the publicity about his money 'is vulgar, boring and generally inaccurate. There may be lots of people that have more money than I have and there certainly are lots of people that have more cash than I have.'

In his old age Getty no longer felt so proud of his money-making abilities and tended to play them down. He now compared himself not to John D. Rockefeller, but to his grandsons, the recipients of the wealth, because 'they, like me, have built their careers on the solid foundations of established and inherited wealth.' When his friend David Rockefeller visited him to recommend that he use his wealth to 'make some significant contribution to the benefit of mankind, such as finding a cure for some terrible disease', his pleas were met with silence. Still Getty told his friends he wanted to be 'the acceptable face of capitalism.'

David Rockefeller was terribly saddened by Getty's decision to leave all his money to the museum. The Rockefellers had long been after Getty to make major contributions to their favourite public projects. In April 1958 John D. Rockefeller III sent historian Alan Nevins's kindly biography of Standard Oil's founder John D. Rockefeller I to Getty in preparation for a request to contribute to the Lincoln Center. Getty's reply was that of a reactionary curmudgeon. He calls the first Rockefeller 'unquestionably the most successful businessman of all time . . . a very good, ethical, kindly generous man'. Getty thought it was unfair to criticize the robber barons of the nineteenth century because that period's business ethics were different from those of the twentieth. Then Getty makes a classic mistake of gaucheness. He tells Rockefeller that his family made a mistake in not policing the faculty of the University of Chicago, which Rockefeller's money founded. Getty advises throwing out all leftists and Communists from the University. After this diatribe he declines to support Lincoln Center on the grounds he is up to the limit of charitable contributions from income and is 'not very keen about giving away capital'. It was only a few months later, however, that Getty drew up his will leaving his entire holding of Getty Oil shares to his museum.

Luckily, Getty never wanted to be a public champion of the oil industry, and as he had physically removed himself to Great Britain he was never asked to testify before congressional committees or to appear on television to defend the profits of the oil companies. But his aversion to paying federal taxes made him the target of some adverse publicity during the presidency of John F. Kennedy. Attorney General Robert Kennedy had discovered that Getty's tax credit in the state of California almost wiped out his federal tax obligation. When Robert told his brother, who had decided to crack down on America's multimillionaires, the President leaked the details of Getty's $500 federal tax bill to *Newsweek*'s White House correspondent, Ben Bradlee, and indicated he would propose a drastic tax reform after the election campaign of 1964. The indiscretion of revealing Getty's tax situation brought immediate complaints to the Treasury and the matter was dropped.

Indeed Getty paid only $503.69 in taxes in 1961 on income of $200,556.36. If he was worried about Kennedy his 1962 tax return did not reflect the fact, for he paid not a single cent of taxes in that year. By 1963, when Getty Oil began to pay cash dividends, thus multiplying his annual income by a factor of nearly 6, to over $1 million, Getty paid taxes of $452,137.11. Apparently his tax advisers had their finger to the political wind in Washington because, when tax reform became a dead issue, Getty returned to his practice of paying little if any taxes. Over the next four years, when his income averaged about $1 million a year, Getty paid no taxes in three of those years.

Tax dodges not withstanding, Getty's name was mentioned from time to time in the 1960s and 1970s as a possible Ambassador to the Court of St James or to France, but his history of dubious sexual liaisons together with his flirtation with the Nazis again prevented him from serving his nation as a diplomat. Each President was deterred by the contents of his voluminous FBI file. Moreover, his fear of flying and ocean voyages prevented him from travelling.

In spite of Getty's claims of friendship, often false, with several US Presidents, the one President he knew well and respected highly was Richard Nixon. Getty ranked him on a par with Roosevelt and found him 'a cheerful, convivial person who enjoyed trading jokes, playing the piano and chatting with friends'. Watergate, he felt, 'was just another minor political scandal'. He had first met Nixon in the early 1960s, after the former Vice-President had lost the presidential race to John F. Kennedy, and he 'encouraged Nixon to return to politics', according to John Pochna, the lawyer who introduced them. When the time came to raise money for political campaigns it is no surprise that Nixon should think of his wealthy friend in England, and so brought him marginally into the Watergate scandal. His aides tried to draw Getty into the network of illegal cash contributions that helped to bring about Nixon's downfall.

The Getty connection began officially on 17 February 1969, just after Nixon entered the White House. H. R. Haldeman wrote a confidential memo to John Erlichman which said 'Bebe Rebozo [a Florida banker and Nixon's closest friend] has been asked by the President to contact J. Paul Getty in London regarding major contributions. Bebe would like advice from you or someone as to how this can legally and technically be handled. The funds should go to some operating entity other than the National Committee so that we can retain full control of their use.'

Erlichman asked an aide to find out if $50,000 gift from Getty for 'social events at the White House' would be 'kosher', which it was not. Then Rebozo asked Herbert Kalmbach, deputy finance chairman of Nixon's campaign, to solicit funds from Getty for the 1970 senatorial campaigns. Kalmbach testified before the Watergate Committee that Rebozo 'set it up for him to see Mr Getty in Europe', but there is no record of a contribution at that time. In 1972, as the re-election campaign gained momentum, Kalmbach asked Rebozo to set up appointments with 'a couple of people that I knew . . . one was Paul Getty and another was Raymond Guest' (US Ambassador to Ireland). This time, according to records kept by the Committee to Re-elect the President, Getty made a contribution of $125,000, which placed him high among the donors to the Republican cause but by no means at the peak. But there was never any proof that the Getty money went into the fund Rebozo maintained in Florida for Nixon administration expenses, or that it was pooled with the $100,000 which Rebozo accepted from Howard Hughes in 1969 and 1970 and which caused such a furore during the Watergate investigation.

Getty was forced to disclose his contacts with the Nixon
administration and reported that Kalmbach visited him in
Italy during the May or June of 1971. He promised to
contribute $50,000 in 1971 and $50,000 in 1972 as long as
it was 'perfectly legal'. Getty also disclosed that he
mentioned to Kalmbach a possible *quid pro quo* – 'that I
might be available for diplomatic duty, if needed. He did
not respond.' The records show, however, that Getty
contributed $3000 each to about fifty different committees
set up across the nation to re-elect the President.

Getty's reward came on 15 December 1972, his eighti-
eth birthday, just after Nixon's landslide victory. First,
the President sent his daughter Tricia, to the birthday
party given in Getty's honour at the Dorchester Hotel in
London. Second, just before dinner was served, a call
came through for the birthday boy from Washington. It
was Richard Nixon wanting to wish 'Happy Birthday' to
one of his cash donors. Getty responded like a little boy,
according to the Duchess of Argyll, who gave the party.
He kept saying over and over, in a tone of astonishment,
'It's the President! The President called me!'

When Nixon's downfall came about, Getty was
emotionally upset. In August 1974 he was with the
Duchess, watching Nixon make his resignation speech.
'There were tears in his eyes,' she says. 'He never thought
Nixon did anything wrong. Even though he was upset, he
asked me to take down a cable to Nixon thanking him for
bringing peace to the nation.' Morally insensitive to
the illegal way Nixon had used his powers, he thought
Watergate was a 'molehill afair' and said Nixon 'would
be a warmly welcomed guest in my home at any time.'
Although Nixon was by no means a dictator, Getty still
was an 'admirer of dictators and of Hitler in particular',
he told his friend, Somerset de Chair.

If he had wanted to attain the doubtful distinction of

being Nixon's largest contributor, he could easily have done so, for his personal wealth was expanding at an exceedingly fast rate. This was due not to some political favour of the Nixon administration but to the explosion of pride, greed and outright blackmail on the part of the Arab oil-producing nations. 'Petroleum is the vital nerve of civilization, without which all it means cannot possibly exist,' President Nasser of Egypt had pointed out way back in 1954, and twenty years later the Arab nations heeded his words: after the Yom Kippur War they imposed an oil embargo on the USA and Western Europe as a protest at the military aid given to Israel. The crisis triggered off the largest and fastest redistribution of wealth in the world's history. The Arabs quadrupled the price of oil from $3 to $12 a barrel in the early 1970s. Getty also got richer, albeit at a slightly less frantic pace than the oil-producing nations. Every time the price of a barrel of crude oil rose $1, the profits of Getty Oil Company rose and Getty Oil shares climbed in the marketplace. Every time the price of Getty Oil rose $1 on the New York Stock Exchange, Getty's personal holding of 4 million shares jumped by $4 million. Likewise, the 7.9 million shares in Getty Oil held by the Sarah Getty Trust gained $7.9 million. By 1975 the Getty Oil Company's profits had more than doubled and Getty was easily able to raise the company's cash dividend from $1.30 to $2.50 a share, thus providing for himself in the last full year of his life an income of the most stupendous size – $25,800,000 – solely from the cash dividends paid to Getty Oil shareholders. Probably no other American entrepreneur had ever collected such a windfall from his own company. The price of a single share, which was well below $100 before the price of oil began to skyrocket, soared to almost $200 a share during 1975. At that moment Getty and the Sarah Getty Trust were together worth just under $2.4 billion.

The maverick oilman identified with the cause of the Arabs who, he said, had been 'gulled, bilked – in a word, suckered. Arabs know that the Western Powers arbitrarily established unrealistically low prices for Arab oil and kept them there for decades.' Similarly he stolidly defended the oil companies' right to make substantially increased profits from OPEC's decisive moves. In the Getty Oil Company's annual report for 1973 he said: 'The basis for the current shortage began building 50 years ago. It was just a question of time. As a nation, regrettably, we have promoted the wasteful use of energy materials, keeping them artificially inexpensive while selling off our inventories at less than replacement costs.'

Such conventional excuses did not help the image of the oil companies. Public opinion polls showed that they were blamed for the energy shortage even more than the Arabs. Luckily for him, Getty was in Sutton Place and not in Washington, where the leaders of the major companies had to appear before one congressional committee after another trying to defend oil prices. But he agreed wholeheartedly with their insistence that the petroleum industry required larger profits as motivation to find more oil and produce it, and predicted that 'the high prices brought about by shortage will make possible greatly increased exploration and production. This, in turn, will increase supply and eventually produce surpluses, which are certain to ease the price picture to some degree.'

Nor had Getty been dozing in his armchair at Sutton Place while the Arabs got their revenge on the industrial nations. Oil had been discovered in the North Sea, and British Petroleum's pioneering exploration which resulted in the Forties field 110 miles from Aberdeen set off a rush by the oil industry to gain a foothold in one of the world's major new energy sources. Naturally, Getty, living so close by, did not want to see the Seven Sisters grab the whole of the North Sea. The British Government also decided that Shell, British Petroleum and Esso, a division of Exxon, the new name for Standard Oil of New Jersey, should not be allowed to monopolize this new resource. However, the cost of exploration and production in this deep water, where the winds and the waves rose to alarmingly dangerous levels, was an obstacle for Getty Oil, which had extensive capital spending requirements in the USA and elsewhere. Because of Getty Oil's failure to discover large new pools of oil, Getty insisted 'that it was better to have a half interest or a quarter interest in a dozen plays than to have all of two or three plays.' In truth, Getty Oil simply wasn't 'big enough to go it alone', says Don Carlos, deputy director of oil exploration at the time.

A formidable joint venture was therefore formed to exploit the North Sea, a combination of extraordinary characters led by Armand Hammer, chairman of Occidental Petroleum. This mysterious entrepreneur had begun his business career in Soviet Russia when Lenin gave him the concession to produce pencils for the USSR. Many years later, in 1958, he acquired a more or less defunct California oil company, Occidental, and annoyed the Seven Sisters by making his own deal in Libya with Colonel Gaddafi. Getty had 'all kinds of misgivings' about Libya and never pushed in person, like Hammer, for an arrangement 'under the right kind of terms', says Don Carlos. Unlike Getty, Hammer wanted to build a diversified industrial empire by acquiring other companies and to become one of American industry's fastest-growing conglomerates. The two American oilmen needed a major British presence on their team, and turned to Getty's friend, Lord Thomson, the Canadian press baron, who owned the London *Times*, the *Sunday Times* and a worldwide chain of newspapers and magazines. To complete the group a US chemical concern, Allied Chemical, was added.

Luckily for them, they did not have to bid at an auction to obtain a foothold in the North Sea. At meetings with British government officials, Hammer was the front man. It helped that he had obtained the largest concession in Libya, a coup for such an independent. Getty sat quietly and was 'very sparing. He was the power behind the scenes. It helped that Thomson was there, a genuine British interest. They made a pretty powerful team,' says John Angus Beckert, then an official of the British Energy Department. 'It was our belief they'd carry out rapid and thorough exploration.' Getty used to invite Beckert down to Sutton Place for a party and then take him into the little study for a private talk. 'He'd endeavour to make you do most of the talking in order to pick your brain. He's a thinker and still waters run very, very deep. He was the most fascinating man, extremely shrewd. He pretended not to hear, but he was listening very carefully,' Beckert told the author.

Occidental owned 36.5 per cent of the consortium, followed by Getty's 23.5 per cent. The remaining 40 per cent was divided between Thomson interests and Allied Chemical. In 1972 they obtained the right to drill in 316,000 acres of water that became known as the Piper field. By the end of 1973 they had discovered at least 642 million barrels of high-quality, low-sulphur crude oil, far better and more saleable than the high-sulphur content oil coming from the Neutral Zone. The very next year at least 400 million additional barrels were found in the Claymore field, 30 miles to the west of Piper. The North Sea finds were timely for Getty since the Saudi

Government forced him in 1973 to reduce production in the Neutral Zone. The consortium built a 30-inch pipeline to bring the oil ashore from the Piper field to the Orkney Islands.

At the age of eighty-three Getty still held the reins over the Getty Oil Company and the lessons he learned early on in Oklahoma and California about the value of the oil in the ground had stood the test of time. He knew the oil in the ground would increase in value over the years. And he had learned that the shares of Getty Oil would be worth more if he could increase the number of barrels of oil that each share represented. By dividing the number of Getty Oil shares into the number of barrels of oil Getty owned, it was found that each share of Getty Oil was worth eighty barrels of oil in 1975. This was far more valuable than the thirty-one barrels represented by each share of Exxon, which was a far larger company in terms of revenues and profits.

Shrewdly, Getty left Skelly Oil independent and never merged the Tulsa-based concern with Getty Oil, even though his opposite number, Bill Skelly, had died in 1957. He liked to have 'two hunters in the field' and estimated that the merger would save barely $500,000 in overheads. It was a simple concept that bore fruit. Skelly Oil, along with two partners, discovered a field in the Persian Gulf worth some 250,000 barrels of oil a day, far better production than that of Getty's Neutral Zone. Skelly also made a promising discovery in the North Sea. By 1975 Getty's Mission Corporation held an investment in Skelly Oil worth $550 million, which had been purchased for $6.4 million.

Naturally, not every venture turned into a success. Getty's $87 million for an Alaskan lease in 1969, a record bid for him at that time, turned out to be a disappointment. The Delaware refinery was still a white elephant, losing money from operations and a costly fire. Getty offered it to Mobil Oil and to Charter Company, but bargained too hard on the price. It must have been exceedingly difficult for him to take a large loss and admit failure on one of his most imperial projects. Perhaps it was his aloofness and distance from the company headquarters and the New York financial world that enabled him to be patient and take the long view. While most other corporate presidents had a few years to make their mark, bank a few million and retire to the country club, Getty was in charge until he died. Among American businessmen Getty was unique because he so completely combined in one place ownership and management. Most other major oil companies were run by hired hands, whose sojourn at the top was brief if well paid.

Sometimes his strategy did not seem to make quick profits. For instance, beginning in the 1960s, he let Aminoil lift more oil from the Neutral Zone than Getty oil took. This strategy meant Getty Oil had a larger share of the oil left in the ground, which Getty felt would be worth more because crude oil prices would advance. When the Arab oil producers began to press for higher prices from the Seven Sisters, Getty began to be proved right. 'This looked bad for individual yearly profits, but in the long run it was the cheapest conceivable form of oil reserve storage imaginable,' says Claus von Bulow, who oversaw this production as Getty's executive assistant. 'From the first day of my association with J.P.G. I found actions or thinking processes which were incomprehensible to senior executives were in fact simply manifestations of his greater experience and superior wisdom. Above all, he was not interested in short-term window dressing of annual profits, but in the long view. The fact that this was not appreciated by security analysts had the coincidental benefit that GOC stock was cheap.'

Getty was apt to seek many opinions about corporate strategy, and for a period in the 1960s he even consulted his son, Gordon, but there was an overall regular rhythm or design to his supervision. Each December the budget team flew to London and worked through each major item of capital expenditure with the old man. Often, the meetings would begin in the late morning and go on until after midnight, with interruptions by art dealers, Getty's women, his regular two-mile walk in the afternoon, and the BBC television news at 9 P.M. Harold Berg, the company's executive vice-president, called him about major investments every week or ten days, but he can hardly remember a call coming in from Sutton Place. Berg's major struggles with the president were over an attempt to institute incentive schemes for employees and to increase corporate donations to the cities where Getty Oil was the largest employer. Getty told Berg 'there was no need for incentives' to work for the Getty Oil Company. Berg changed his terminology to a bonus plan, which Getty then accepted. All his life he felt that donating the company's money was 'stealing from stockholders', meaning himself, of course. It was the only time that Berg ever got into trouble. The old man believed his $17 million gift of the museum was charity enough. Berg's tenaciousness finally won out, and Getty Oil began to join the corporate-responsibility movement in the 1970s, but as a corporate citizen the company undoubtedly had its shortcomings since it reflected the image of the president and chief stockholder.

What Getty still loved best were his financial dealings. He personally managed the negotiations with Saudi Arabia for permission to delay paying Getty Oil's tax bill for one year. This meant that Getty Oil had the use of revenues ranging from between $150 million and $500 million to finance its activities, an advantage no other oil company was accorded in the Middle East.

Getty was a shrewd negotiator. He could stay up working far later than anyone else, without food or drink. Often he never offered sustenance to others in the room, with the result that they were dying to get the matter over and leave to get some refreshment. He also intimidated his opponents by means of a fierce concentration and an absolute, unbending need to prevail.

By the end of his life, Getty had more personal notoriety than any other oilman. Dr Hammer had built a more broadly based industrial empire but personally owned very little of it. The Getty name adorned thousands of red and orange signs at gasoline stations across the USA. Getty television commercials bombarded every American household. Exxon commercials, by comparison, never mentioned the name Rockefeller.

The possibility of selling the whole of Getty Oil occurred to him late in life. The Shah of Iran, through the National Iranian Oil Company, once made an informal bid, according to Giorgio Schanzer, a close friend of Getty. But Getty wisely could not figure out what to do with the money, since the Sarah Getty Trust limited him to reinvesting the proceeds in the bonds of a specified number of nations. The reinvestment would fare badly during the high inflation of the 1970s. Moreover, Getty absolutely refused to pay a 33⅓ per cent capital gains tax on the enormous profit he had created. Getty would die the 'richest American', never giving up control and steadily along the way getting rid of other shareholders who might impede him in his quest, whatever that might be.

20
The Mortal

I never enjoyed making money, never started out to make a lot of it. Money doesn't necessarily have any connection with happiness. Maybe with unhappiness.

J. Paul Getty

Sutton Place was meant to be a family manor house, not a museum or an oil company headquarters. Even though there was a constant flow of guests, including oilmen, titled nobility, art curators, old girlfriends, Getty's life in the rose-brick Elizabethan home, set in its vast green parkland, seems isolated and unhappy, even if he insisted that it was not. 'He didn't quite know what pleasure was,' said Jeannette Constable-Maxwell, who had known him since she was sixteen. There was no shortage of business appointments, formal dinners and visits to galleries or museums. Yet he felt like an orphan, or so he said. This suggests that, no matter how busy he was at eighty-three, he had never recovered from the lack of the warm intimacy children are supposed to feel for their parents. Many strangers who came to Sutton Place were intimidated by his strange power of concentration, the deadly cold stare. If he did not intimidate in person, then the Alsatians or the roaring lions caged near the swimming pool were there to do it for him. They made him seem fiercer than he really was.

The atmosphere of Sutton Place was forbidding also. Only the soft light and the mellow wall panelling were warm. He never turned on the boilers, so that guests had to wear sweaters and sometimes even overcoats in their rooms. 'It was like a tomb,' said one of the upstairs maids. He told his guests the cold 'was good for the art', proving once again that he cared more for his possessions than he did for people. Mrs Joseph Thomas, the lively Swiss wife of a Getty Oil board member, once took a walk in the garden all by herself and then could not get back in the house as all the doors were locked. And still in the 1970s there was no phone in her room and she was forced to use the payphone. At dinner, she, her husband and Getty sat alone at the long table that had seating for several dozen people. After dinner the dining-room doors were locked so that no one could return. 'I never wanted to go back there again. It was too dreadful,' she says.

'When I think of Paul, I think of money' was the comment not of an enemy but of his friend, the Duke of Bedford. For more than six decades Getty had accumulated money and oil with a single-minded devotion that was awesome, even compared with his collection of art, wives, lovers and sons. Perhaps it was his failure to make human connections, to appreciate the needs of others, that forced him to accumulate money.

The money gave him freedom, gave him power, gave him the ability to be as eccentric and bizarre as he liked – and get away with it. But did it make him happy? Some close aides said he was a happy man because he could control his life more than most men. C. Lansing Hays Jr, his lawyer and executor, found him to be 'a delightful, charming, intelligent man who could tell hilarious jokes.' He had a marvellous ability to mimic other accents and used to explain to men friends how he decided the amount of money to be given each lady. In a southern black dialect he would say, 'When I'se soft, I'se hard – and when I'se hard, I'se soft.' His son Gordon could have been describing a happy man when he said his father 'was a fathomless man, commanding and disarming, philosopher and clown. He was inscrutable, a showman, a prince of players. He was charismatic and even mesmeric . . . He was stoical in grief and at last jocular on the day he died.' This does not sound like a miserable human being, although it is perhaps a rather overdrawn portrait, full of denial.

Gail Getty, his ex-daughter-in-law, always felt there were several separate Gettys. There was the businessman, hard, cold and demanding. Then there was the sentimental Getty who sat in European restaurants until late at night spinning tales of the old days in the Oklahoma and California oil patch. Gail found him to be most romantic when he was in love, especially during the late 1950s with Mary Teissier. He also tried to be a parental figure, always admonishing Paul Jr and others to act properly. Gail felt this was the 'Victorian Big Paul'. At times other than the kidnapping 'Big Paul could be incredibly sweet to me and the children,' she says.

The money he amassed caused him a great deal of trouble. Every relationship was overshadowed by it. Almost everyone he met wanted some of it, or so he thought. 'The rich are not born skeptical or cynical. They are made that way by events, circumstances – and most especially by the countless people who have barracuda leers (and barracuda instincts and intentions) hidden behind their broad and beatific smiles,' he wrote at the end of his autobiography. 'I cannot help but think that while there is much, very much, that money can buy or make possible, there are not enough millions in the world to make every wish and dream come true.'

One wish that did not come true was a Getty dynasty. George II, the heir apparent, was dead, a 'probable suicide'. Paul Jr had a history of drug problems and was abandoned by his father. Gordon decidedly was not a businessman. Ronald, the only active businessman, did not show any great skill at making money. Indeed, his Getty Financial Corporation, a miniconglomerate of restaurants, real estate and oil and gas holdings, was in the red steadily from 1973 to 1976. Only Getty himself had the Midas touch. Rather than a dynasty, all Getty accomplished by his genius was to make two sons, Paul Jr and Gordon, automatic millionaires every year from the trust's income – and to cause Ronald grief over his $3000. His sons' failings only served to illuminate more vividly their limitations and his genius at creating wealth.

He could have retired to become a beachcomber, as he fantasized from time to time, but he claimed there was no one to step into his shoes. That was a piece of self-deception by a man who had made sure there was no dynasty, no son who could replace him. He even claimed on one occasion that he lacked control over his life. 'I never had control over things. My life has really been like that of the soldiers in the Light Brigade. Theirs not to reason why – theirs but to do and die.'

Still, his friend Brealey felt, 'He was the happiest man I ever met. He was so proud of what he achieved. People like to pity the rich, but in this case it is not true.' Brealey always puzzled how this yokel became the world's richest man. In the end, he decided that an 'extraordinary shrewdness lay underneath the coldness, cruelty and naïveté'. He understood that Getty succeeded because he 'never lost his cynicism about people's behaviour and that he saw excitement or emotion as a weakness. On these two scores Getty had iron discipline,' says Brealey.

There were times when he seemed to crave companionship, but if he did he was harking back to his ambivalent relationship with his parents. He never seems to have got over their loss. 'I used to see orphans and feel sorry for them because they had no parents,' he said in 1975 when

he was already eighty-two. 'Now I have no parents and
haven't had for many years. I was very close to my
parents. I used to rely on my parents. I've had several
wives but only one mother. I never realized I'd be an
orphan one day.' On Thanksgiving night that year, the
taste of his rum and Coca-Cola nightcap made him think
of the roast turkey and cranberry sauce of his childhood.
He had to overcome the nostalgic pull of that past and
the solution was readily available. 'When I go upstairs to
my bedroom, my thoughts are on the work I will have to
do tomorrow,' he said.

He needed women for company, entertainment and
sexual pleasure. He liked variety and could afford it, but
the craving for variety was the sign of the restless traveller, the nomad who moved from city to city without
having any roots. 'There was not such a woman born who
could cope with all [the] requirements of his special social
position, his interests, his knowledge and, at the same
time, with his very simple way of personal life and family
life,' says Anna Hladka. He used to complain to her of
'the misery of his life'. 'He cried. He was unhappy with
all the women making trouble.' He told her, 'I will die
before you. I will be waiting for you on the other side of
life. Then we can start a nice, simple, happy life.'

The trouble was that his women friends wanted to
organize his life, boss him around, meddle in his affairs.
Penelope was for ever telling him how to handle the
servants and whom to invite for dinner. Mary always had
an opinion on the art collection, even on the oil business.
They sensed that he needed some controlling influence, a
mother. That is why they meddled, some uncontrollably,
and why, according to Marianne von Alvensleben, their
relationship became 'a nightmare'. But he showed her 'a
kindness that was unique. Right from the beginning there
was utmost respect.'

Getty, in fact, never had a solid loving relationship with anyone. He thought a lasting relationship with a woman was only possible if you were a 'business failure'. That was his smokescreen. If you were a business success (and who was a greater success?), then expect divorce. He really could not help himself. Actually, he was a failure at marriage because he did not know how to love. He knew two chief ways to relate, sexually and by means of a financial transaction. There were a few women, like Jeannette Constable-Maxwell, who were simply devoted friends, who enjoyed the simple teaparties and the walks at Sutton Place without the subject of money or sex coming up.

'Money was a curse on him, the idea that distorted everything he did,' says Mary Teissier. Perhaps so, but this did not deter the women including Mary, from bringing the subject up. They would go constantly to his secretary, Barbara Wallace, when they wanted something, whether it was a car, a house, a journey, something for the children or just money.

'It's not the money, although I'm sure that's part of it,' said Rosabella Burch who was close to him at the end of his life. She spoke of 'a sort of physical presence which holds them . . . Women all want to be near him. Most of his friends are women. They all want to marry him.' She wanted to do so herself. 'I love him,' she told the press in 1976, when she was forty-three years old. 'I want to marry him.' Rosabella lived at Sutton Place around 1970–71, and then Getty bought her a small house at nearby Cranleigh.

As he became older and weaker he still contrived to exercise a kind of bizarre tyranny by keeping everyone in a state of expectancy over his will. The intrigue and scheming at Sutton Place reached alarming proportions. The king's vassals were in a constant state of tension, trying to be close to him, to please him – and to inherit. He played his role to the hilt, getting perverse pleasure from the grovelling.

To the very end his women friends vied with each other to mother him. For his part, he had always preferred travelling with several women at the same time. It meant variety, and he did not have to pay attention to any single one of them. In his diary he makes continual references to the women closest to him, using his pet names for them – Pen for Penelope, Ros for Rosabella, U for Ursula and Mary for Mary Teissier. U drives him everywhere in her Mercedes 450. He takes walks with Mary or goes to see Pen's new kitchen. On Christmas Day they all give gifts to each other as though they are part of one big happy family. There is no sign in his diary of irritation over their demands; he merely catalogues the simple things he does each day with the different women, just as he reported in his wilder days in Berlin during the 1930s.

He probably preferred Penelope, because she was most like him – very cool, disciplined and calculating, immensely polite, and terribly well turned out. She was his most trusted social pilot and kept exact records of every plant and tree at Sutton Place, the kind of service that appealed to Getty.

However, when Penelope in 1973 decided to marry again, Getty became wild with anger and threatened to cut her out of his will entirely. Once before, in 1969, he had reduced her bequest dramatically, but had restored it a month later. This time he could not bring himself to treat Penelope so severely. On 20 March 1973 the sole purpose of the fifteenth codicil was to reduce her 7500 shares to 5000.

According to Robina Lund, Getty had become totally cynical about women by the end. He said to her, 'The majority of very wealthy single men will tell you sadly that most, though not all, women can be divided into two types, those that you pay to stay with you and those that you pay to stay away.' This hostility for women was confirmed just after his death by his publisher and friend, Mark Goulden, who called him 'a lecher, a miser, a womanizer whose private life was often bizarre beyond belief. He never conquered his fear of death and he tried in vain to stay the ravages of time by frequent facelifts. He was secretly contemptuous of the sycophants who surrounded him, particularly the members of his harem. He lacked affection, yet he yearned to be loved. Throughout his adult life, love eluded him.' Every time a woman asserted her independence, Getty struck back. When Mary sailed off to Egypt on one of his oil tankers, he expressed his displeasure by moving her to a bedroom far from his.

Of all the women, Robina Lund, who had been one of the nearest and dearest, had farther to fall than most. Her financial reward was reduced from 6250 shares of Getty Oil to 1000 shares and then to none at all, while her monthly stipend was cut to the peculiarly insulting amount of $209. According to Bramlett, who kept his master's accounts in all these private dealings, 'She wanted to marry. She was trying to organize his life, and he didn't like to be organized. There was some incident, perhaps a banquet or charity affair, where he embarrassed her and she got angry. As a result he took her out of his will.' At the end, she was handling his public relations and was paid at a rate of about $20 an hour.

Just as some women fell out of favour, new ladies began to make their appearance in the will. In October 1973, during the kidnapping of J. Paul III, he bequeathed to Lady Ursula d'Abo 1000 shares of Getty Oil stock, which placed her in third position behind Penelope and

Mary. However, he did not include her in the list of women to receive monthly cash dispensations. Two years later, in 1975, he added Rosabella Burch, who was named for 500 shares, only half Lady d'Abo's legacy.

Nine women were favoured in the will with amounts ranging from a mere $100 a month for Hildegard Kuhn, whom he had met in a Berlin dance hall as long ago as 1930, to Penelope Kitson's $1167 a month. The purchasing power of even these bequests fell sharply after he died, as, he had neglected to make any provision for inflation. As we have seen so often, his parsimony amounted almost to a sickness. He told a German woman friend, Ingeborg Weger, who sent him the gift of a pocket watch, that HM Customs had recommended refusal of the parcel on account of the customs duty. Getty asked her to send the watch to Marianne von Alvensleben who would bring it to Sutton Place on her next trip in order to avoid import duty.

In the last months of his life the scenes at Sutton Place became even more bizarre. Once Getty was in the middle of discussions with Harold Berg, his executive vice-president, when Ursula d'Abo appeared, as she apparently often did, uninvited. She knocked on the window and said, 'Paul, I have to see you.' Exasperated, he said, 'I can't see you. Go away.' Another night, at dinner, Mary Teissier started an argument because there was no bidet in her bathroom. She said American women were 'filthy beasts' because they didn't use bidets. Teddy, his fifth wife, who was on her last visit, turned to him and said, 'Paul, we had a happy married life without a bidet.' Teddy, who had divorced him in 1958, claims that in late 1975 Getty proposed marriage once again. There is no rational explanation for this, unless he recognized some desperate need to be attached to someone. His greatest pleasure at the close of 1975 was to invite a group of his women admirers to Sutton Place for dinner on the same evening and then watch their reactions. He would retreat to his television set at 9 P.M. for the news, enjoying the spectacle of ladies vying for his favours.

Mark Goulden felt that Getty 'had a certain innate nobility of character that could of blossomed under the loving care of a selfless woman devoted to the task of humanizing a man who, without such help, was destined to become a money machine and little else.'

Getty may have been ambivalent about the women and his sons, but he apparently loved to spend time with his favourite art historian, the brilliant Frederico Zeri. They would sit up late at night reading Catullus or take long walks around Sutton Place conversing about history, art and literature. Zeri, a thickset Italian of Turkish descent, first studied botany at the University of Rome and later became a noted expert on Italian paintings of the Renaissance. He believed Getty to be a greater man than Berenson. 'Paul was so sheltered,' Zeri said, 'a mystery, a man who needed affection, because he had made a mess of his private life.' Zeri loved Getty, even though he found him to be 'a Puritan [who] is always the one who sins, believing only others are sinners. He whitewashes himself.'

Even Getty's preference for inanimate objects and animals, neither of whom could talk back, to humans was characterized by a lack of trust. His insatiable obsession to know every minute detail of an object's provenance was his way of learning to have faith in his ownership of that object. As for humans, von Bulow always felt that 'Paul read and read and read. He read enough so a man or a woman couldn't bluff him.'

Schanzer believes, 'Paul never recovered from George's death. It was the beginning of the end for him. He lost the will to live. He lost the heir to the throne.' Still,

Getty's desire for the figment of a dynasty caused him to reinstate Gordon to partial grace in 1973 by making him a co-trustee of the trust. Then, in 1975, the nineteenth codicil to the will was written, adding Gordon to Ronald as an executor of the estate. This change, it was felt by those close to Getty, was a result of the ministrations by Gordon's wife Ann, who kept in close contact both with Getty and the staff at Sutton Place during the last years of his life.

Even as the end approached, Getty could not reconcile himself to its inevitability. He continued to take every precaution that was available and chewed each morsel of food many times. He persisted with injections of H-3, the painkiller Procaine, which is banned in the USA because of the claim that it restores youthful energy. He seems not to have worried about the possible side effects of Sinemet – the involuntary and abnormal movements of the head, mouth or limbs – hoping that the pills which he took every day would reduce the tremors and the sluggish movements of his body. Tom Smith, a giant oil engineer who claimed to have American Indian blood, was often called in to massage Getty's feet and shoulders, a treatment that was supposed to have a magical curative effect.

Getty took to telling his friends that his fondest wish was to return to sunny California. The ranch house behind the museum was prepared for his arrival, incorporating all the usual security measures such as the ordering of fencing and steel reinforcement for his bedroom door. Leon Turrou travelled to California and wrote a detailed list of instructions, including the admonition that the front door should not be opened. Preparations were made to send Getty across the Atlantic Ocean on a liner, and then by transcontinental railroad, surrounded by bodyguards, to California. He also wrote to Armand Hammer asking to be flown home on the Occidental Petroleum chairman's private jet. 'I am a resident of Great Britain and I think I am technically a resident of Italy. I am domiciled in California,' he had testified some years earlier. 'I am temporarily sojourning in England and intend to return to California when my work is done,' he said. But he never did so.

Guarded by a man with a shotgun, the aged billionaire took his afternoon walks around Sutton Place. A chute had been attached to his bedroom window so that he could slide to the ground in case of danger. Inside the bedroom he had half a dozen police truncheons to protect himself against interlopers. But first an intruder would have to get past the guard in the corridor, and Rebel, the fierce successor to Shaun, Getty's favourite dog.

The last time Getty left Sutton Place on business was on 22 January 1976 for a meeting at Lord Thomson's office about the North Sea fields. 'He left the meeting early because of muscular pains. Dr Mackenzie, his personal physician, told him not to worry, that he would live a long time. By 7 February, however, he felt severe discomfort in his left hip muscle and required the attentions of Tom Smith, who pressed on a nerve and put a muscle back in place. At the time Mackenzie concluded that Getty's problem was fibrositis or muscular rheumatism caused, in part, by his sleeping in an adjustable chair in his study. Instead of the best surgeons from London, Getty was visited by an acupuncturist who inserted needles in the top of his head, between his eyes, in his painful shoulder, and two needles in each heel. Nevertheless, his discomfort did not prevent him from approving the purchase of several art objects for the museum, including an amphora for $70,000. On 11 April he notes in his diary an unusual occurrence – a nightmare 'about the hidden staircase. Wish I had taken it out.' He does not spell out the dream sequence, but he equates the dream with the real hidden staircase in Sutton Place. Finally, on Easter Sunday, 18 April, Getty asked Mackenzie to obtain a second medical opinion. On 20 April Mackenzie brought a Dr Simmons, who told Getty that there was 'no problem, other than my age'.

Finally, in early May, Mackenzie took a blood test and on the 5th gave Getty the 'final results of the blood test and their conclusion'. Nothing more is revealed in his diary about the apparent news of a fatal illness. There is no emotion whatsoever, not even a hint that he knows that he is dying. In the end it was an internal disease that captured Getty, cancer of the prostate. When Dr Mackenzie told him the news, Getty asked Barbara Wallace to fetch the encyclopedia and open it at the page that began with 'Ca'. Methodically, he read all the scientific details of his disease as if he were looking up the provenance of a painting or the log of an oil well. But on 11 May he reports that 'U' visited him for an hour, 'after which I felt quite nervous'.

Dr Mackenzie wanted him to go to hospital for an operation because 'most men over eighty have cancer of the prostate and it doesn't usually kill them'. Getty adamantly refused. He was afraid of anaesthetics because they would make him lose control. How sad that such a wealthy man, who prided himself on his depth of knowledge in so many fields, did not send for the greatest medical experts in Britain and have a thorough investigation to determine what was wrong with him. 'He preferred it that way,' says Barbara Wallace. 'He didn't want to go into hospital and have all the things done. He was scared. He had never been in hospital.'

Dr Mackenzie told him there was another possible treatment for the cancer, pills that contained female hormones which might be able to 'contain' the growth of the cancerous male hormones. During this treatment, the doctor explained, a man's breasts often increase in size. Getty agreed to take the tablets, but kept asking Dr Mackenzie the all-important question: 'Will it affect my sex life?'

He remained in his study, sleeping in the electrically operated chair that folded down into a bed, even though a lift had been installed so that he could go upstairs to his double king-sized bed covered in green damask. 'He wanted to die with his boots on,' said Harold Berg. The phone in the study was disconnected so that he could be undisturbed.

Just before he died he borrowed almost $900,000, even though he held $10.5 million in short-term Treasury bills and another $30 million of liquid investments, as if to prove his claim that he had no cash. Even more amazingly, late in life he appeared to lose faith in the shares of Getty Oil and began to accumulate a portfolio of shares in other oil companies, his competitors, Standard Oil of California, Union Oil, even Exxon, Mobil, Texaco, Gulf, Arco, Shell, Continental, Amerada Hess and Occidental Petroleum. Is it possible that, even at that age, he invested $28 million in order to keep a weather eye on the competition?

The man who had not been able to face his son Timmy's illness did not want people viewing his own illness either. Just a day or two before he died a former mistress appeared, freshly coiffed and smartly dressed, to pay her last respects. She leaned over Getty and called out to him. 'Oh Paul, Paul dearest. I love you. What do you have to say to me?' The dying man tried to raise himself from the chair that was serving as his deathbed, but failed. He strained to make words, but all that came out was a deep groaning noise. And then, for a split second, his old furious will restored itself. 'Get that woman out of here,' he said as forcefully as he could and slumped back.

Gordon was sent for, but when he arrived, Getty was livid, according to several reports. He did not want to see his son. Ronald was in Cape Town, South Africa, where he was living at the time. Paul Jr, totally estranged from his father, was living close at hand in Cheyne Walk.

Even though a nurse was in attendance, it was Norris Bramlett who had to help the dying man to urinate and empty out the basins. One night, which Dr Mackenzie thought might be the last, Barbara Wallace sat up the entire night holding his hand after Penelope, Gordon, Norris and the doctor had gone to bed. In the morning he woke and said to her, 'I'd like my breakfast.' Later that day he walked around the library.

Right at the end Ann Getty flew in from San Francisco and asked if he would like to be flown back to California. He still seemed unable to make up his mind. 'He couldn't go to hospital, and he couldn't fly,' says Bramlett.

For the last few days his faithful friend, Jeannette Constable-Maxwell, stayed at Sutton Place, sleeping outside the study. While he was awake she read to him from his favourite Henty books, a mother figure some fifty years his junior, reading him boys' stories. The last book Jeannette read to him was *Beric the Briton*, a novel about Roman rule in Britain at the time of Nero. Henty had written it in 1893, the year after Getty's birth. Beric, who becomes governor of the British province, declares that there are two Neros: 'One . . . gently inclined to clemency; desirous of the good of his people, and of popularity; a lover of beautiful things; passionately devoted to art in all its branches; taking far greater pleasure in the society of a few intimate friends than in state pageants and ceremonies. There is another Nero; of him I will not talk. I desire, above all things, not to know him.'

Getty obviously had some affection for Nero because he named the lion cub that Jeannette had sent him as a birthday present several years earlier after the Roman Emperor. Nero was 'the handsomest and biggest lion, an absolutely magnificent lion,' she says today, and Getty was 'prepared to talk to him because he trusted the lion more than humans.'

Nero, the last of the Caesars, was a symbol for Getty. He too was more than one person. The Roman Emperor seemed to be virtuous and generous. At other times he became detestable, feared and hated. He lusted after his mother and then had her put to death. He executed his wife in a sadistic way, and took other lovers, male and female, as he saw fit. But Nero also squandered his fortune, wasting money on an obscene palace, which he then burned down, along with vast portions of Rome.

Getty's diary reveals certain elements of his character at the very end of his life. He admits shame at not remembering the anniversary of Shaun's death, the dog whose loss had supposedly kept him in his room for three days. Apart from that, he mentions the pains in his shoulder, the number of times he walked across his study, the number of hours he slept, who telephoned that day – nothing more. The major events in his life are the purchase of another painting or a piece of furniture. To the end, however, he received the constant attention of Penelope, Mary, Ursula and Rosabella, who took turns in caring for him. They were in attendance like his nurses. The last entry is for Wednesday, 2 June: 'U here for two hours, Pen here.'

Some superstitious members of his entourage found deep meaning in the timing of Getty's death. When he breathed his last in the early hours of 6 June 1976, it was exactly three years to the day since his heir apparent, George II, died. The same people believed that a gypsy fortune-teller had warned Getty not to return to California at the risk of death. But he told Robina Lund that a fortune-teller had once told him he would die on foreign soil. 'I don't want that to happen,' he said. 'I want to go home, however ill I am – I'll even fly to get there in time.' He had felt all along 'a stranger in an alien land – nothing can ever replace the country of your birth.'

And so, on Tuesday, 8 June, an English oak coffin, lined and upholstered in swansdown and quilted silk, was delivered to Heathrow Airport in time for TWA flight 761. Destination Los Angeles. The richest American was going home.

Epilogue: The Aftermath

Even in death Getty was a nomad. When his body arrived in Los Angeles the coffin was found to be too small and cheaply made, like the coffin of a 'Western outlaw', according to Evey who was responsible for the burial arrangements in the USA. The billionaire was placed in an expensive bronze coffin. However, Getty's request to be buried near his $17 million museum could not at first be granted because burial on private property is against the law in California. As a result, for two and a half years, his body was temporarily moved from place to place in the Forest Lawn cemetery, where many Hollywood greats have found their resting place, until the Getty lawyers, led by Edward Landry of the Los Angeles firm Musick, Peeler & Garrett, managed to persuade the city planning commission, the zoning office and the coastal commission that Getty should be allowed to lie under a stark slab of grey granite on a slim promontory jutting out into the Pacific, undisturbed by the hordes visiting his art collection. There he is guarded by closed-circuit television cameras which monitor the movements of intruders.

In death, as in life, Getty's wealth was the target of demands from disgruntled lady friends, members of the family, government tax collectors, even perfect strangers. His probate file fills several vast volumes containing a multitude of requests for payment and a precise inventory of every object he owned, together with its estimated value. The tradition of a fragmented family – who share very little besides Getty's eccentric genes – is the back-drop for a soap opera that has more money and perhaps more human tragedy than *Dallas* and *Dynasty* put together. The spirit of alienation has resulted in a decade of bitter litigation that has set brother against brother, son against father, niece against uncle. All the suits somehow represent Getty's legacy – the emotional baggage that goes along with the money. Gail Getty is sceptical that the Getty superwealth has brought any happiness. 'Big Paul created this legacy – and what it's done, it's created unbelievable unhappiness. People have grown up who don't know how to cope with this, who they are, the name. If he did anything that's really wrong, that's it.'

J. Paul Getty, 'a symbol of oil, wealth and power', according to the *New York Times*, left two separate pools of wealth together worth more than $2 billion when he died in the early hours of Sunday, 6 June 1976. His personal holdings in Getty Oil Company, some 4 million shares, were worth exactly $661,943,577.50. This fortune was to be handed over as a tax-free gift to the J. Paul Getty Museum in Malibu, California. The Sarah Getty Trust, with almost 8 million shares of Getty Oil stock, was worth about $1.3 billion. These shares were part of an even grander tax-avoidance scheme. They were not intended for his three surviving sons, Ronald, Paul Jr and Gordon. This gift was to skip an entire generation and be handed over to Getty's sixteen grandchildren decades later. The income from the $1.3 billion, however – some $20 million in Getty Oil dividends – was to go to Paul Jr, Gordon and George II's three daughters, Anne, Claire and Caroline.

The British Inland Revenue tried to claim that he had changed his domicile to England in order to tax the estate. Her Majesty's Government did not really push hard enough, but the estate lawyers were frightened. Witnesses were called upon from around the globe to testify that not a week went by without Getty's returning to Malibu to discuss business. Von Bulow offered to testify that his patron was 'one of the last representatives of a great Puritan tradition, whose high standards and sense of duty' kept him in the oil business rather than beside his swimming pool in Malibu. The truth is that Getty died just in time, because his accountants had told him that Capital Transfer Tax, which was introduced in 1975, would begin to apply to him in April 1976. Getty's claim to be resident in California was his last great ruse, all for the grand purpose of limiting his tax bill. Like most spectacularly rich men, he had an obsession about not paying taxes.

In addition, Getty had an investment portfolio, bank accounts and real estate valued at over $60 million. He liked to claim, moreover, that the collection of furniture, sculpture and paintings in the Getty Museum was worth another $200 million. He also left some unutilized assets: a membership of the Santa Monica Beach Club worth $1000 and twenty-five days' accrued holiday pay worth $19,097.25. His personal diaries, filled with the minutiae of his daily life, were valued at $1. He died with $16.41 on his person.

Getty's cardinal rule for disposing of his estate was to give nothing to the government because they wasted it. He very nearly succeeded in this goal. Secondly, he wanted to prevent his children and grandchildren from obtaining vast wealth without working for it. On this score he failed. His third goal was to leave a monument – the J. Paul Getty Museum. On this point he achieved immortality. And his fourth rule, although he did not state it publicly, may very well have been to leave friends and retainers the bare minimum.

The Getty Oil shares he left to various women were worth less than 1 per cent of his holdings. He left monthly stipends to a slightly larger group of women at a total cost of $636,000 in invested capital. There were outraged screams from near and far. Rosabella Burch, who got $82,000 in Getty Oil shares, demanded $500,000 and settled for $150,000. Anna Hladka was given $17,500 to educate her sons George and Karol. Getty's ex-wives got short shrift. When he died he was still supporting four of them with monthly cheques. Jeanette received $250 a month; Fini got $3515.24 a month (this comparatively large sum included a cost-of-living increase); Ann received $1000 a month; and Teddy, the longest wed, had financial support of $8333 a month. After he was gone, however, there was little they could do to him, so he left nothing to Jeanette, Fini and Ann, while cutting Teddy down to $4583 a month. Teddy found it difficult to believe what he had done.

Getty's closest aides were not happy either. Barbara Wallace received a year's salary, which amounted to little more than $5000. Bramlett, Getty's alter ego, got six months' pay or $36,000. One of the aides told his friends that Getty had promised explicitly or implictly to take care of everyone; and added that there was only one word to describe his manner of using them – 'Greed.'

The first Getty to sue was his granddaughter Anne, one of George II's three daughters, who charged that the twenty-first codicil, signed in March 1976 three months before he died, was a conspiracy by Getty's aides Bramlett and Hays to control the money going to the museum. This codicil diverted Getty's personal wealth into a trust which was to oversee the museum and would not be controlled by the family. The court decided that Anne's $1.8 million annual income from the Sarah Getty Trust and the promise of an eventual share then worth $123 million were sufficient compensation for bearing the Getty name.

The second Getty to sue was Ronald, who had the greatest cause to be bitter at his treatment. He still received only $3000 a year from the Sarah Getty Trust, while his half-brothers, Paul Jr and Gordon, received cheques of $1,231,878.68 every three months. In his will Getty left Ronald only 2000 shares of Getty Oil worth $320,000. Ronald felt humiliated, since his mother, Fini, had told him that he would be put on the same footing as Paul Jr and Gordon. It is true that, as his father's executor, Ronald received fees totalling $4.2 million, but he was terribly shaken to learn that his father had not kept his longstanding promise to leave him a legacy. He decided to sue both the Sarah Getty Trust and the Getty Museum to obtain redress.

His lawsuit caused extra anxiety to the museum, which could not receive the bulk of the estate until it was settled. What is more, if Ronald could overturn the will, his victory might cause revocation of the tax-free gift. The Internal Revenue Service, using Ronald's claims, alleged that Getty's gift of shares was not a charitable contribution and demanded $628,631,523.81. 'We were in danger of losing the whole inheritance,' says Edward Landry, the tax lawyer responsible for the estate. The museum settled the suit by giving Ronald a $10 million tax-free present. This gift cost the Getty estate over $30 million in taxes, but it was still a far cry from the $628 million the government wanted. 'We were gambling $10 million against $600 million,' says Landry.

But Ronald was not satisfied with the $10 million. He still wanted to be treated equally with his half-brothers in the Sarah Getty Trust. Gordon offered Ronald $1 million to settle the suit, hardly a generous proposition. In order to win, Ronald needed to prove that his father had defrauded him in 1934, when the trust was first established, by giving him only $3000 a year. This was because Fini had refused to divorce him at a time when Getty desperately needed to marry Ann in order to legitimize Paul Jr. Ronald's lawyers suggested that Paul Jr and Gordon should voluntarily relinquish a share of their annual income, but, once again, the estrangement of the Getty's prevented any settlement. It was not until January 1984 that a Superior Court judge in Los Angeles ruled that Ronald had no case. He was doomed to receive only $3000 a year until he died.

By comparison, the wealth left Paul Jr, Gordon and George's three daughters with an unimaginable cornucopia of ever-increasing income for which they did not have to lift a finger. As Getty Oil raised its cash dividend from $1 a share in 1978 to $1.40 a share in 1979 to $1.90 a share in 1980 to $2.30 in 1981 and $2.60 in 1982 – an annual growth rate of almost 25 per cent – the income of these five Gettys skyrocketed because they received 100 per cent of the income from the Sarah Getty Trust. As a result, by the early 1980s Paul Jr and Gordon were collecting some $28 million a year – almost the same amount their father earned in 1975, the last full year of his life. It is one of the few times in financial history that the sons of the wealth creator had more money to spend than the wealth creator himself.

Getty could not have suspected that his fourth son Gordon would play a role in doubling the family fortune by causing the Getty Oil Company to disappear. Gordon, whose vast income from the trust allowed him to dabble in music, took his father's place on the board of Getty Oil in 1976. He also was one of the three trustees of the Sarah Getty Trust, along with the lawyer Hays and the Security Pacific Bank in Los Angeles. However, the bank was frightened by the prospect of lawsuits against the trust, especially by the British Inland Revenue, and decided to withdraw. As a result, Hays and Gordon shared the trustees' fees, which, at 5 per cent of the trust's income, amounted to $1,650,000 per year.

Hays not only dominated Gordon completely but became a thorn in the side of the management of Getty Oil. He tried to become J. Paul Getty's successor and often clashed with the professional oilmen in Los Angeles. In 1977 the final step of amalgamating the Getty oil interests took place when Skelly Oil and Mission Corporation were consolidated with Getty Oil. The merger gave Getty Oil a ranking in the largest fifty US corporations for the first time. In 1980 an offer by the Kuwaiti Government to buy Getty Oil was rejected. The oil company, benefiting from a substantial flow of cash, began to diversify into insurance and cable television, a move which Getty undeniably would have opposed. It was no surprise that Gordon did not approve of this diversification either.

Great relief was felt at Getty Oil when Hays died in May 1982. The company thought it would now have a free hand, but it had not reckoned on Gordon coming into his own as the sole trustee of the Sarah Getty Trust's 40.2 per cent block of Getty Oil stock. Ignored and belittled by Getty Oil's professional management, Gordon attempted to assert himself and win recognition as his father's rightful heir. The management and directors were not astute students of Getty financial history. They should have realized that Gordon, the only son to confront his father in court, might not tolerate their blatant disregard. Sidney Petersen, chairman and chief executive officer of Getty Oil, was unable to disguise his low opinion of Gordon as a businessman. Gordon told his best friend, Bill Newsom, that he was fed up being treated as a nonentity by the Getty board. Moreover, he was disturbed that Getty Oil shares had fallen to the low $50s by the end of 1982, a third of the asset value ascribed to Getty Oil by Wall Street. 'Had Petersen mollified Gordon at the outset, Gordon might have been as happy as hell,' says Evey. 'He just wanted to be massaged by the management of the company.'

Gordon never realized how similar he was to his father. He understood that his father 'had a chip on his shoulder and worried about people laughing at him', but there was one major difference between the son and the father. Gordon was incredibly generous, especially to family members who were in trouble. When his brother Paul Jr refused to pay medical expenses for Paul III, Gordon stepped in and paid them. He supported Martine and the children. There was no sign of a mean streak unless you counted his inability to give voluntarily part of his income to Ronald. Gordon knew that Ronald had been treated unfairly but could not bring himself to change the situation.

Like all Getty's sons, Gordon adopted his father's attitudes. In mid-1982, shortly after Hays's death, he began examining alternative courses of action to increase the value of Getty shares. He was partial to a plan to buy in Getty stock and retire it, so that the Sarah Getty Trust would increase its holdings and voting power from 40.2 per cent to over 50 per cent, just as his father had once done. But the Getty management did not want Gordon to have a majority interest and be able to rule at will. The company would agree to buy the stock, they told Gordon, but the trust's voting power must remain at 40.2 per cent.

Unlike his father, Gordon was neither an oilman nor a financier. He was unfamiliar with the machinations of the corporate takeover field and could be fooled by the scheming of his opponents on the Getty Oil board and influenced by far more sophisticated investment bankers on Wall Street. He himself never had any concrete propositions to make to the board. 'He was all pie in the sky as far as I was concerned,' says Berg, who liked Gordon but nevertheless was part of the plot against him.

Gordon needed an ally and proposed to join forces with the J. Paul Getty Museum which still owned 11.8 per cent of Getty Oil after selling some shares in the market. Harold Williams, chairman of the museum and a former business executive, would not join Gordon in this scheme. Nevertheless, the threat that the museum might back Gordon led the Getty Oil management and directors to develop a secret plan to thwart him. At a hastily called board meeting in Philadelphia on 2 October 1983, which Gordon did not attend, Petersen told the board of a plan to ask another member of the Getty family to petition the court for an addition of the Bank of America as co-trustee. This move, if successful, would dilute Gordon's voting power. The company had already approached Paul Jr through his London lawyer, Vanni Treves, about blocking Gordon. Paul Jr, uneasy about his business ability, said 'Monopoly was always Gordon's favourite game at fifteen, but we didn't have world-class competition then.'

In London on 5 October Gordon, the museum and the company came to an apparent truce. A 'standstill' agreement, drawn up by Martin Lipton, a shrewd New York lawyer representing the museum, seemed to make peace. Williams, for the museum, would become a Getty board member, but no Getty shares would be repurchased.

This move set off a wave of speculation in Getty Oil shares on the New York Stock Exchange. Wily traders sensed that Getty Oil might be 'in play' – meaning that the internal struggle between the family and the company might result in the sale of the entire company. Ivan Boesky, a major investment manager who specialized in takeover candidates, acquired a large holding in Getty Oil because he felt 'a rumbling of discontent in the breast of a man in [Gordon] Getty's position was enough to set off a chain of events he himself would not be able to control. I believe the son wanted so badly to get back at the father.'

Boesky was correct that the peace would not last long. By 19 October, at a meeting in San Francisco, Gordon decided to press for additional representation on the Getty board. He was given four additional seats and influenced, it appears, by his wife Ann chose the impressive financial muscle of Laurence Tisch, chairman of the Loews Corporation, Alfred Taubman, a Detroit property developer, who had just taken over Sotheby Parke Bernet, and Graham Allison, dean of Harvard's Kennedy School of Government. Another nominee, Warren Buffett, a brilliant investor who lived in Omaha, Nebraska, was unable to take the assignment.

The Getty Oil Company accepted the changed voting configuration of the board but pressed forward with its plan to dilute Gordon's sole voting power over the trust's 40.2 per cent interest. At a board meeting on 11 November in Houston Gordon asked to leave the room. The board then discussed their secret plan to block him. With the co-operation of Paul Jr, a highly reputable Los Angeles attorney, Seth Hufstedler, representing Paul Jr's fifteen-year-old-son Tara Gabriel Galaxy Gramaphone, was prepared to petition the court for a co-trustee to be appointed. Gordon did not have the foggiest notion that he was being bamboozled. 'Perhaps I'm naïve and gullible,' he said. 'I was raised to resist the notion of a plot.' Gordon has that curious ability to deny reality that was so evident in his father's political thinking. 'The oil business is one of the most honest in the world,' he said recently. 'A man's handshake is as good as his word.'

Open warfare between Gordon and Getty Oil was now declared. Lipton, the museum's legal grey eminence, shot off a letter to Williams and Petersen charging that the standstill agreement had been violated. In Gordon's camp the choice of Tara as petitioner was regarded as an act of the deepest cynicism. Tara, who lives with his maternal grandparents in France and attends a private school in Dorset, had seen his father not more than twenty times since his mother died in 1971. According to Mark Getty, Tara's half-brother, Tara spends perhaps twenty minutes in his father's presence about once a year. At the time Tara was chosen as the tool to block Gordon, he had not seen his father for at least six months and had no knowledge that a lawsuit was about to be filed in his name in California. It was Hufstedler who brought the lawsuit in his name as the boy's guardian. The filing of Tara's suit on 15 November forced the museum into an alliance with Gordon which was the worst possible outcome for the company.

Now Lipton wheeled out the Delaware corporate law which enables holders of over 50 per cent of a corporation to dismiss its officers and directors. Clearly Getty Oil was in a state of chaos and vulnerable to a takeover. Into this battle stepped Pennzoil, a medium-sized oil company, eager to get hold of Getty's Oil's valuable domestic oil reserves. The Pennzoil intervention served Gordon's needs beautifully. Pennzoil proposed to pay Getty Oil's shareholders, other than the trust, $110 a share in cash. If this transaction went through the trust would end up holding an even greater proportion of Getty Oil. Gordon would become chairman, while J. Hugh Liedtke,

Pennzoil's chief executive, would be named president and chief executive.

Even as the Getty board was pushing Pennzoil to raise its offer to $112.50 a share, Petersen was secretly looking for another way to foil Gordon. The investment bank, Goldman, Sachs, was requested to find a 'white knight'. Unluckily for Gordon, the Pennzoil deal was held up by a last-minute lawsuit brought by his niece Claire Getty Perry, one of George II's three daughters. Claire was properly worried how the trust would be valued in the future, since Gordon was turning Getty Oil back into a private corporation without any publicly listed stock. Also, the terms of the trust were supposed to prohibit Gordon from selling Getty Oil shares unless there was a possibility of a 'substantial loss'. So there was a very real legal problem involved.

Earlier Claire had asked Gordon why the trust, then worth $1.8 billion, needed to be increased in value. Gordon's answer made little concrete sense. 'A very interesting philosophical question,' he said. 'It is my fiduciary duty to maximize the wealth and income of the trust.' If this was an accurate reflection of his innermost thoughts, then Gordon could achieve this goal in only one way, by the sale of Getty Oil to the highest bidder. In fact, while Gordon was intending to become chairman of Getty Oil in tandem with Pennzoil, his financial adviser, Martin Siegel of Kidder, Peabody & Co., felt it might be necessary to liquidate the company after a year's time.

While Claire's lawsuit was being heard, Goldman, Sachs came up with a bidder – Texaco, one of the Seven Sisters that Getty had always opposed. The first step in Texaco's strategy was to offer Lipton, representing the museum, $125 a share, an offer too good to be turned down. That same night in early January 1984, the same offer was made to Gordon in his Pierre Hotel suite. By accepting the offer Gordon doubled the value of the Sarah Getty Trust from $2 billion to around $4 billion. Within a period of eighteen months he had inadvertently made more money faster than his father had ever done.

Siegel, Gordon's investment banker, felt his client was 'a man intent on doing the right thing. He wanted to maximize value for the shareholders more than he wanted to be chairman. Economics drove the decisions he made rather than personal aggrandizement.' Siegel believes Gordon was smart to be a buyer of Getty Oil at $110 a share, a wholesale price, and to be a seller over $120, at Texaco's bid, a retail price. For their services Siegel's firm, Kidder, Peabody, received $15 million. The Getty Museum was a seller too, because it had to reduce its 11.8 per cent interest to 2 per cent by 1992 for tax reasons.

With Texaco's offer the Getty clan became aroused again. This time the interveners were Ronald's children, who went to court arguing that Gordon had not obtained the maximum value for the company. Texaco immediately raised its bid to $128 a share, giving a value to Getty Oil Company of $10 billion, at that time the largest corporate acquisition in American history.

Gordon was hailed as a financial genius for this record transaction, but some family members and aides agreed with Landry, who was 'convinced Mr Getty never wanted the company to disappear. It was the single thing outside of the museum in which he had pride . . . The mistake Mr Getty made was not naming the president of the oil company to be executor of the trust. That would have identified the interest of the trust together with the company for ever,' says Landry. However, oil prices were soon to drop and the value obtained by Gordon makes the deal in retrospect look like a brilliant coup. Bramlett felt that 'Gordon's done the family a great favour. The family had its wealth in the stock of a badly run, medium-sized oil company. The oil business is a precarious one. The trust doesn't dissolve for several decades, and there might not even be an oil company left. Gordon locked in the wealth. Now there is a guarantee of great wealth.'

Still, it is unlikely J. Paul Getty would have approved the sale of his company to Texaco. After the transaction Texaco fired more than half Getty Oil's 19,600 employees. The jobs Getty had so proudly created no longer existed. Even the distinctive red and orange Getty signs at service stations were scheduled to disappear. Inside the company the level of anger at Gordon rose to new heights.

Now that Gordon was sole trustee of a $4 billion trust, he became, according to *Forbes* magazine, 'the richest American', just like his father. It was not strictly true, since the $4 billion was in trust for the next generation. But Johnny Carson, on his *Tonight* show, joked, 'How would anyone begin to spend $4 billion?'

At this point the battlefield shifted from corporation *versus* Gordon to Gordon's relatives *versus* Gordon. The suit by Tara attempting to have the Bank of America appointed as co-trustee was still in effect. Gordon was also obliged to inform all Getty beneficiaries of any 'non-routine transaction which significantly affects the trust estate . . . at least five days before entering into any such transaction.' The $4 billion was placed in US Treasury bonds which, at that point, produced an income of more than $1 million a day.

Gordon then took an important step that was to inflame George's three daughters and raise the temperature of legal warfare among the Gettys to an even more intense heat. On 27 February 1984 he filed a secret petition in the Los Angeles court naming four trustees to succeed him in the event of his death: his wife Ann, Gail Getty

(Paul Jr's former wife), Ronald and Gloria Getty (George II's first wife). It is hard to believe that he thought such a move would make peace in the family. 'Gordon behaved irresponsibly,' said Treves, Paul Jr's lawyer. 'This move brought the nieces out of the woodwork.' On 19 June the three nieces filed suit claiming that Gordon had violated 'at least the spirit if not the letter' of the court's order. Their suit asked for Gordon to be removed as a trustee and to pay the entire tax on the sale of Getty Oil shares to Texaco. This lawsuit went on for almost a year, providing substantial fees for a handful of California law firms. Gordon was subjected to three weeks of questioning which was videotaped under floodlights in his attorney's office. 'It was psychodrama,' says Edward M. Stadum, attorney for Paul Jr's children. 'His nieces were trying to punish Gordon within the bounds of the law.' The three girls – termed the Georgettes by the lawyers – were embittered by the notion that had their father lived he would have been in charge of the trust as well as the oil company.

Even as the lawsuit proceeded, so did negotiations for a settlement. Gordon was by now fed up with the protracted aggravation. A lengthy court trial promised to unveil every dirty little secret of the Getty family history. The solution seemed to be to divide up the trust into four separate new trusts – one for Ronald's family, another for Paul Jr and his children, a third on behalf of George's three daughters, and a fourth for Gordon and his children. Gordon would voluntarily relinquish his 5 per cent fee on the trust's income, which would now go to the trustees appointed by each separate family group. In the case of Paul Jr, the solution became a method of financing his five children, including Mark, a twenty-four-year-old charmer with the red hair of the clan's founder, and Paul

III, who would now be able to pay his own way and support his family.

However, Gordon's lawyer would not let him split the trust into four parts unless he was given an indemnity against future losses of capital. On the eve of the court battle Bill Newsom, now a State Appellate Court judge, helped introduce legislation in California making it possible to divide an existing trust into several parts. He wanted to make peace among the family and to provide a solution that would give an assured income to Paul Jr's children, including his own godson, Paul III.

After the Gettys pay $1 billion in taxes on the sale to Texaco, four separate pots of $750 million each will fall under the principal jurisdiction of Paul Jr, Gordon and Ronald, and Anne, Caroline and Claire. Gordon is waiving his 5 per cent trustee fee on the income from the trust so that others, including some nieces and nephews without an income, can receive its largesse.

Long before these legal battles were resolved another sordid and tragic series of lawsuits were proceeding involving Paul Jr's children. In 1976 they were forced to sue their father because he was not setting aside a portion of his income in trust for them. In his 1966 divorce from Gail, Paul Jr had promised to deposit in trust for his children 15 per cent of his income over $54,000 and 10 per cent of everything over $154,000. Apparently, he had put aside nothing of his $1.3 million income in 1975. This was at a time when Paul Jr's health was breaking down, largely as a result of the life he had been leading. Seldom if ever has there been a similar case where a very rich man has perceived in the nick of time the perils of unlimited wealth, turned his back on the past and transformed himself into a benevolent philanthropist as Paul Jr was to do.

Since the kidnapping, Paul III's life had been a series of

mishaps and difficulties. In 1975 his maternal grandfather, Judge George B. Harris, filed suit to become his guardian because, according to the record, Paul III was 'financially improvident' and 'unable economically to care for his own concerns'. In the same year he was also arrested in Malibu for stealing a pickup truck after he had crashed his own car into a guard rail. He was released on $2000 bail. Soon after his grandfather's death he was arrested in London's Hyde Park for 'behaving in a manner likely to offend public decency' and was fined £10.

But a worse tragedy awaited Paul III. In 1981 he was in a 'serious and debilitating medical condition' caused in part by his 'overconsumption of alcohol', according to a lawsuit he brought in 1984. During his treatment, according to his deposition, he was given several drugs, including Placidyl, a powerful depressant, Valium and Dalmane, which are also reputed to have a 'powerful depressant effect', and Methadone, a drug which is used to help people reduce dependence on more serious drugs like heroin and cocaine. On about 5 April 1981 he collapsed in to a coma and suffered permanent brain damage. When he regained consciousness he was almost totally blind, could not speak except to make vague guttural sounds, and was paralysed. He required round-the-clock medical supervision which cost $25,000 a month.

At first Paul Jr refused to pay the medical expenses and the boy and his family were forced to go to court in an effort to obtain the money. Paul Jr argued that because he lived in London a California court had no jurisdiction over him, a position that caused outrage on the part of most of the other Gettys and the presiding judge. Only after Paul Jr's personal doctor had travelled to California to see the boy for himself did the father make the money available. In the meantime, Gordon paid for the medical expenses and helped finance Paul III's family.

On 27 June 1984 Paul III brought a charge of medical malpractice and drug product liability against Abbott Laboratories, the maker of Placidyl, Hoffman–La Roche Inc. and Roche Products, the makers of Valium and Dalmane, as well as against several doctors and two medical centres in southern California. The suit is still pending.

The division of the Getty fortune will secure the emotionally wounded Gettys from a brutal world and provide generous trust funds to enable every one of them to live a life of leisure if he or she so wishes. The days of creating a large fortune are now over to all intents and purposes. It is a new era in which the utilization of the stupendous income has become the dominant theme, just as it has for the J. Paul Getty Museum.

Ironically, in his will Getty created the richest museum in the world and put it in a position to accomplish what in his lifetime the richest man in the world had failed to do. The Getty Museum, or rather the trust which administers the museum, had an endowment of over $2 billion in June 1985. Because of the law affecting charitable trusts, it must spend between $100 million and $120 million each year, or about 4.25 per cent of the market value of its endowment. Just as ironically, the man who sought absolute control over his affairs left his monument a totally free hand to use the money as it wished. It was an absolutely amazing and uncharacteristic legacy. Getty placed no restriction on the money so long as it was used 'for the diffusion of artistic and general knowledge'. The museum's trustees can sell every work of art and collect early maps or postage stamps if they so wish. By way of comparison, Henry Clay Frick, whose museum Getty greatly admired, prohibited his executors from buying or selling a single work of art. 'For the first time in history here was a man who left unlimited funds for a museum –

a fascinating and exciting idea' said Peter Wilson, the late chairman of Sotheby's. Getty's bequest stunned the museum's curators because he had kept them on a tight rein in his lifetime by insisting that the endowment was limited to the original $55 million.

Brealey felt the bequest to the museum 'showed great wisdom and generosity of spirit. Dead hands should not rule from the grave.' He thinks that the gift of this money will remove the East Coast's cultural disparagement of the West Coast. 'It will put California on the map.'

It is unfortunate that Getty never came under the influence of Duveen for more than a very brief period. There was never any dealer or art historian who could dominate Getty and change his natural character, which was ruled by the need to obtain a bargain. Getty was similar to the millionaires of the Duveen era only by his inner drive to become a lord or emperor by acquiring works of art of earlier centuries. Berenson's fatherly advice to him was followed from time to time, but in a limited fashion. He did not have either the vitality and exuberance of Duveen or the scholarship of Berenson, nor was he a tiger like Norton Simon who wanted to collect the best, whatever the cost. Getty probably knew that Frick's entry in the *Encyclopaedia Britannica* devoted more space to his collecting art than to his violent squashing of the steel strike. The only way Getty was bound to achieve the status of a Morgan or a Mellon was to leave more money to his museum than anyone else had ever done before.

Grandiose works lie ahead for the Getty Museum. A vast art centre, the largest cultural project in southern California, will be built over the next several years outside the city of Los Angeles. The antiquities will remain in Malibu, but the vastly expanded collection of paintings and decorative arts and the newly acquired prints and photographs will be moved to the new centre. The Getty Museum has its eye on achieving another splendid coup – the acquisition of the industrialist Norton Simon's unequalled collection of paintings. It is the collection that Getty himself should have made had he been willing to spend money and press forward aggressively by careful steps. Jennifer Jones, the former actress and Simon's wife, has been appointed to the Getty board. Two paintings, Degas's *L'Attente* and Poussin's *Holy Family*, have been bought jointly and will hang for six months of each year in each museum. Otherwise there is no precise strategy for accomplishing the consolidation of the two collections. 'We are interested in the collection and with Jennifer Jones on the board that means in close rapprochement,' says Gordon, a director of the museum.

The Getty Museum has emphasized that it does not wish to monopolize the world market for masterpieces or drive prices higher than a work's true value. Despite its attempts at good citizenship, the feeling of envy that used to be directed at Getty is now attached to his museum. This antipathy became headline news when Paul Jr, in the summer of 1984, contributed $500,000 to a national campaign to keep the fourteenth-century painting *The Crucifixion of Jesus* by the Sienese master Duccio in Great Britain. 'I was fed up with everything streaming to Malibu. It's time somebody stopped it,' said Paul Jr, who had been thrown off the board of the museum years earlier.

Unhappily, Paul Jr's earlier life has left him a physical wreck. He has such severe circulatory problems in his legs that he can walk only a few yards without great difficulty, according to his surgeon Lionel Gracey. He also has impaired lung function as a result of an attack of pneumonia and irreversible cirrhosis of the liver, together with signs of impending diabetes. Over the past five years he has been admitted nine times to the London Clinic and twice to the Wellington Hospital for several months at a time. In fact, his doctors now prefer that he remain in the London Clinic. The bottle of rum a day he used to drink has been replaced by bottles of beer kept in a refrigerator. He is no longer the slim, striking blade-about-town. Today, bearded, generally unkempt and some 15 pounds overweight, he is no longer a swashbuckling ladies' man. But he has the wealth and ambition to become in his own right a broadly based philanthropist like John D. Rockefeller.

Paul Jr has the unparalleled opportunity to give vent to his artistic sensibility. 'I have much greater income than I shall ever need,' he says. 'The most important thing I have to do in life is to see that this income is used to the best possible purpose.' In June 1985 Paul Jr announced a £50 million gift to the National Gallery with no strings attached. He does not mean it to be a competitive response to the Getty Museum. Unlike his father, he wants to repay the nation that has been his home for the past twenty years. He is not constantly planning to return to California.

As a result of Paul Jr's gift, William Waldegrave, Under-Secretary of State for the Environment and a government spokesman on the arts, declared in the House of Commons, 'Maecenas has come amongst us,' a remark which drew cheers from the House. Britain does not yet realize that this Getty is likely to become the nation's greatest philanthropist. Paul Jr plans to establish another £50 million trust which will donate money for education and social causes in Britain. He has given well over $8 million to the British Film Institute to build the nation's first film library. In addition, he has changed his will to leave his entire estate to charitable projects now that his children are taken care of by the Sarah Getty Trust.

His brother Gordon also strives to express the artistic sensibility inherited from his father. He studied music briefly in 1960 and after a hiatus of eighteen years began to compose again and give singing concerts. His song cycle based on the poems of Emily Dickinson – *The White Election* – has been performed several times, in part financed by himself, to mixed reviews. The themes of Emily Dickinson, a recluse from her late twenties, were nature, death and hopeless love, and strike similar chords to those of the Getty family. Gordon's second performed work is *Plump Jack*, a musical portrayal of Falstaff. Gordon was ecstatic at the positive reviews he received after its premiere. He prefers the civilized and beautiful solitude of his music study in his San Francisco town house that looks out on the shimmery bay. There he can peacefully compose his music and listen to his extensive record collection. Unlike Paul Jr, he is a devoted family man, but is often teased for his absent-mindedness and his avoidance of complicated emotional situations.

Gordon's wife Ann is the one Getty fully involved in the world. Her intelligence, vivacity and money are serving to turn her into the most well-known member of the Getty family. Tall and striking in the latest fashions and sometimes ostentatious jewellery, she has become a leading hostess of private parties and charitable functions in New York and San Francisco.

Ann Getty, better than all the family members born with the Getty name, is capable of spending the money for a wide variety of causes. She will undoubtedly contribute part of the family's wealth to the institutions of which she is a leading member. She has become a trustee of the Metropolitan Museum of Art, a director of Sothebys Parke Bernet and a trustee of New York University. Last winter, along with her close friend Alexander Papamarkou, she took a hundred people on a boat trip down the Nile, importing $75,000's worth of French wines for the occasion. It was one of the most talked-about happenings of the 1985 international social season – Arab royalty mixed with Greek shippers, French designers and American jet-setters. In early August, she rented the elegant Cipriani Hotel in Venice for an international conclave on opera, one of her husband's passions.

Her ambition to be at the centre of intellectual life in both London and New York was fulfilled during 1985 by the purchase of an interest in the prominent British book publishing company Weidenfeld & Nicolson. In partnership with Lord Weidenfeld, she has bought Grove Press, an avant-garde American publishing house which has published Jean Genet, Samuel Beckett and William Burroughs, writers that her late father-in-law would have abhorred. Ann's visibility has become widespread. Her picture appears in the society pages and the fashion magazines regularly. She has recently acquired an apartment in London, in a fashionable building in Chelsea also inhabited by Lord Weidenfeld. The Gettys mean to spend half the year in their new, elegant Fifth Avenue apartment, just as Getty himself did during the 1930s.

Ann and Gordon Getty are the new American royalty, one of the most sought-after couples in society. The only media appearance they have made together was an interview by Barbara Walters on ABC television's *20/20* programme. Ann refused to admit to 30 million viewers that to be rich means one has to be unhappy. 'I think it's possible to be very rich and happy and, I suppose, very poor and happy. But it's easier to be very rich and happy, I think.' Gordon, for his part, guaranteed that someday he would be known as 'Gordon Getty, composer' rather than 'the richest man in America.'

The only businessman among this generation of Gettys is Ronald, who refused to discuss the affairs of his company, Getty Financial Corporation. However, he has not made a resounding success by all reports. In 1976 the Securities and Exchange Commission suspended trading in shares of his corporation. Like his father, Ronald is afraid of flying, but is reputed to be a good family man and has maintained an extremely low profile.

Anne, Claire and Caroline Getty, the so-called Georgettes, have adamantly refused to be interviewed or have their pictures taken. Their lives as members of the family are entirely private. The husband of one Georgette told an interviewer, 'We are private people and trying to remain that way. People hate rich people.'

After the payment of taxes the Sarah Getty Trust will have about $3 billion in assets, which means that each one of Getty's grandchildren is worth on paper in June 1985 almost $200 million. Assuming prudent investment decisions are made with this fortune, one day the *Forbes* list of the wealthiest individuals in America will include more Gettys than all the Rockefellers, Mellons, Kennedys and DuPonts put together.

Appendix I
Getty's Annual Income and Federal Income Tax Paid, 1923–75

During the 1920s, 1930s and 1940s Getty was able to use the oil depletion allowance, equal to 27 per cent of exploration costs, and intangible drilling expenses as a deduction from his income. While these book-keeping entries cost him no outlay of money, they reduced his taxes by an enormous amount. In 1929, for example, the year the stock market crashed, Getty earned $281,085.80, mainly from individually owned oil and gas wells in California. But he paid tax of only $12,046.82, less than 5 per cent on his income. In 1932, the worst year of the Depression, Getty paid tax of $5290.18 on income of $76,371.13.

In the late 1940s Getty began contributing art objects to the Los Angeles County Museum to help shelter his income. As his personal income grew during the 1950s and 1960s Getty began to use the donations to the tax-free Getty Museum at his Malibu ranch to reduce his tax burden. Every year he would donate works of art valued at a level fixed for him by friends in the art world. Often the Internal Revenue Service would question the values placed on the furniture and paintings, setting off a protracted controversy. Getty was constantly fighting with the tax authorities and for many years protested the California state income tax.

By the early 1970s Getty's donations of art had to be substantial in order to shelter his stupendous income. In

1972 Getty donated $9.7 million to the Getty Museum, and in 1975 he gave $14.7 million. These donations were either in the form of shares of Getty Oil, the works of art themselves or shares of Art Properties Inc., a small company which owned the works of art. Another company, Fine Arts Corporation, owned the objects at Sutton Place and Palo as a means of legally holding them in the USA and preventing the British and Italian tax authorities from ever possessing them in lieu of taxes.

The use of art donations worked beautifully as a tax dodge. In 1975, the last full year of his life, Getty earned over $29 million, but paid tax of only slightly over $4 million, a rate of 14 per cent. Not bad for a billionaire.

To give an idea of the value of Getty's income over the years in present terms, the following figures may be helpful:

$31,000 in 1923 = $195,000 today
$281,000 in 1929 = $1,750,000 today
$119,000 in 1934 = $950,000 today
$1,297,000 in 1948 = $8,750,000 today
$661,000 in 1957 = $2.5 million today

	Total income ($)	Federal income tax ($)
1923	30, 961.95	0
1924	4,897.57	0
1925	249,149.46	42,986.04
1926	133,213.47	0
1927	103,371.90	3,863.84
1928	121,974.11	19,676.16
1929	281,085.80	12,046.82
1930	55,106.45	0
1931	56,822.03	3,023.35
1932	76,371.13	5,290.18
1933	104,612.50	9,667.90

1934	118,621.27	5,152.23
1935	106,235.64	4,790.52
1936	122,172.26	258.54
1937	268,357.95	147,791.04
1938	57,968.01	8,208.20
1939	missing	missing
1940	192,537.22	86,267.42
1941	243,631.57	36,346.66
1942	197,261.15	49,451.00
1943	314,009.21	15,495.15
1944	403,714.35	192,763.43
1945	128,371.57	744.14
1946	110,544.24	7,380.30
1947	598,905.27	227,405.11
1948	1,297,443.13	281,008.52
1949	152,184.68	5,948.92
1950	268,533.01	59,164.62
1951	1,972,013.77	732,902.66
1952	missing	missing
1953	missing	missing
1954	690,474.65	251,510.75
1955	1,030,333.53	371,913.01
1956	764,500.00	281,352.32
1957	661,465.72	265,279.18
1958	261,915.93	126,773.17
1959	118,886.70	19,263.88
1960	201,783.17	32,530.76
1961	200,556.36	503.69
1962	190,309.11	0
1963	1,177,493.41	452,137.11
1964	1,182,746.70	0
1965	1,246,438.86	19,857.25
1966	1,089,391.86	0
1967	781,898.64	0
1968	missing	missing

1969	missing	missing
1970	12,982,429.00	1,848,106.00
1971	missing	missing
1972	14,815,471.00	3,298,866.00
1973	16,301,554.00	missing
1974	18,355,835.00	missing
1975	29,467,308.00	4,200,180.00

Appendix II
Chronology of the Getty Fortune

1892   George F. Getty worth $150.000, J. Paul is born.

February 1916   A $500 investment in Nancy Taylor lease garners J. P. Getty $11,850 profit.

June 1916   J. P. Getty claims his first million from dealing in Oklahoma leases.

January 1917 George F. Getty's net worth is $1,621,590.14.

1923   J. Paul Getty claims his $1 million has grown to $2.3 million through expansion in the California oilfields.

December 1928   J. P. Getty buys a one third interest in George F. Getty Inc. for $1 million ($250,000 cash, $750,000 in notes).

October 1929   J. P. Getty watches the stock market crash in New York.

May 1930   George F. Getty dies, leaving J. Paul only $500,000 of his $10 million estate. Sarah Getty controls two thirds of George F. Getty Inc.

September 1930   J. Paul begins to buy shares of Pacific Western Oil using margin account at E. F. Hutton & Co.

December 1931   Iraq rejects J. Paul's bid for an oil concession.

February 1932   J. Paul realizes $4.5 million by selling Pacific Western's Kettleman Hills oil acreage to Shell Oil.

15 March 1932   J. Paul begins buying shares of Tide Water Oil at $2.50 per share, an all-time low. The $4.5

million from the sale of oil properties is used to purchase more shares of Tide Water Associated.

Christmas 1933 Sarah Getty sells the remaining two thirds of George F. Getty Inc. to J. Paul for $4.5 million. He now is in 100 per cent control of George F. Getty Inc. J. Paul gives his mother no cash, only notes for the full amount.

Christmas 1934 The Sarah Getty Trust begins with $3,368,000 in its coffers. Sarah contributes $2.5 million in notes owed her. J. Paul contributes $868,000 in the value of stock. He pays not a cent to his mother.

1935 J. Paul exchanges the trust's notes for shares in Getty Oil enterprises and an option on 300,000 shares of Tide Water Associated at $11 per share.

January 1935 J. Paul begins buying shares of Mission Corporation which owns 14 per cent of Tide Water Associated and 55 per cent of Skelly Oil Co. By spring 1935 Tide Water shares have risen from $2.50 to $11 per share.

January 1937 J. Paul controls 40 per cent of Mission Corporation. The price of Mission shares has almost tripled, from $10.50 in January 1935 to $29.50 in January 1937. Tide Water Associated shares sell at $20, up from $2.50 in March 1932. In comparison, the Dow Jones industrial average has doubled since 1932.

December 1938 Sarah Getty Trust is worth $18.573,489 – an increase in value of six times in four years. George F. Getty Inc., 100 per cent owned by J. Paul, is worth $27,353,977. Therefore, the Getty fortune is valued at $45,927,466.

April 1939 J. Paul buys the Pierre Hotel for $2,350,000. It had been built in 1930 for $10 million.

May 1940 Sarah Getty challenges the irrevocability of the trust. Pacific Western shares are selling around $7 each.

May 1946 J. Paul merges George F. Getty Inc. into Pacific Western, which becomes the pinnacle of the Getty oil empire. As a result, J. Paul owns 457,227 shares of Pacific Western, or 34.3 per cent. The Sarah Getty Trust owns 699,422 shares, or 51 per cent. (The average price for Pacific Western is $26.44 in 1946.) Together, the Gettys owns 1,156,649 shares of Pacific Western worth approximately $30,581,799.

January 1949 J. Paul acquires oil concession in the Neutral Zone for $9.5 million. In addition, he owes $1 million per year plus a 55c per barrel royalty.

March 1953 The first oil is found in the Neutral Zone.

1953–8 J. Paul invests $600 million in refineries, tankers and other expansion.

1956 Pacific Western changes its name to Getty Oil Company.

October 1957 At the age of sixty-five, J. Paul is named the richest American by *Fortune* magazine. His fortune is estimated at $700 million to $1 billion. J. Paul and the Sarah Getty Trust together own 11.3 million shares of Getty Oil (81.14 per cent, worth some $300 million). The market value of J. Paul's shares in Tide Water, Mission Corporation, Skelly Oil and Mission Development brings the fortune to $700 million. At the time, forty-five Americans are worth more than $100 million each.

1958 DeGolyer and McNaughton estimates that the Neutral Zone holds 12 billion barrels of oil.

1959 Mandatory oil import quotas damage Getty's plans for expansion. J. Paul decides to shrink the empire.

1965 Sale of Tidewater's western properties to Phillips Petroleum nets $309 million.

1967 J. Paul merges Tidewater into Getty Oil. He remains president of Getty Oil. George F. Getty II is executive vice-president. As a result, J. Paul personally together with the Sarah Getty Trust controls 12,573,789

shares of Getty Oil or 62.24 per cent, worth $968 million
(11 August 1967).

1968 *Fortune* magazine declares that J. Paul Getty and
Howard Hughes are tied for the honour of richest American at $2 billion each. Again *Fortune* has combined the
stock market value of Getty's holdings in Getty Oil,
Mission Oil and Mission Development.

1972 North Sea leases acquired in consortium with Occidental Petroleum, Thomson Newspapers and Allied Corporation. By 1974 the consortium has found 1 billion
barrels of crude oil. During this two-year period, OPEC
quadruples the price of crude oil.

1974 The roughly 12 million shares of Getty Oil owned
by Getty and the Sarah Getty Trust are worth $1.9 billion.
During 1974 J. Paul personally received dividends from
Getty Oil stock amounting to $15,600,000. By compari-
son, eighty-four members of the Rockefeller family held
assets of $1.3 billion.

January 1974 The J. Paul Getty Museum is built in
Malibu at a cost of $17 million.

6 June 1976 J. Paul Getty dies, leaving 4 million shares
of Getty Oil worth $661,943,577.50 to the museum. The
Sarah Getty Trust, owning almost 8 million shares of
Getty Oil, is worth $1.3 billion.

January 1984 The Getty Oil Company is sold to Texaco
for $10 billion. As a result, the Sarah Getty Trust is
worth $4 billion, producing income of over $1 million per
day. The J. Paul Getty Museum endowment is worth
$2.2 billion, making it the richest museum in the world.
Gordon Getty, sole trustee of the Sarah Getty Trust, is
named the richest American by *Forbes* magazine.

Between 6 June 1976 and 30 April 1984, the income
created by the Sarah Getty Trust came to
$443,748,323.71. One third of this revenue went to
Gordon, one third to J. Paul Getty Jr and one third was

divided between Anne C. Getty, Claire E. Getty and
Caroline Getty. J. Ronald Getty received $24,000 from
the trust or $3000 annually.

April 1985   The Gettys agree to divide up the $4 billion
Sarah Getty Trust. They have paid over $1 billion in
capital gains to the US Treasury and the state of California, the largest individual tax payment ever collected in
that state. Someday, when Paul Jr, Gordon and Ronald
die, the sixteen Getty grandchildren will each receive an
equal share of approximately $200 million at its 1985
valuation.

Appendix III
The Twenty-One Codicils to Getty's Will, 1958–76

Conditions of the Original Will, 22 September 1958

Hal Seymour (cousin)	$5000
Ethel LeVane	$500 per month for life
Margaret Feuersaenger	$125 per month for life
Hildegarde Kuhn	$75 per month for life
J. Ronald Getty	10,000 shares Getty Oil
J. Paul Getty Jr	8000 shares Getty Oil
Gordon Getty	8000 shares Getty Oil
Louise Lynch Getty	$55,000 per year for life
J. Paul Getty Museum	Remainder of the estate

Any person declared by court action to be a wife or child $10,000

Executors: David Hecht (lawyer), George F. Getty II, J. Paul Getty Jr (a.k.a. Eugene Getty), J. Ronald Getty

Codicil No. 1, 18 June 1960

Penelope Ann Kitson   2500 shares Getty Oil

Codicil No. 2, 4 November 1962

	Shares	*Monthly allotment*
Ethel LeVane		$250
Penelope Ann Kitson		$500
Mary Teissier	1000	$400

Codicil No. 3, 20 December 1962

	Shares	*Monthly allotment*
Margaret Feuersaenger		$125
Hildegard Kuhn		$75

Penelope Ann Kitson		$500
Mary Teissier	1000	$400

Executors: George F. Getty II (two votes), J. Ronald
Getty and J. Paul Getty Jr. Gordon Peter Getty's name
is crossed out.

Codicil No. 4, 15 January 1963

	Shares	Monthly allotment
Margaret Feuersaenger		$200
Hildegarde Kuhn		$100
Marie Teissier	2000	$500
Penelope Kitson	3500	$750
Robina Lund	1000	
Gloria Bigelow	500	
Mary Maginnes	500	
Belene Clifford	500	
Karin Mannhardt	200	

Executors: George F. Getty II, J. Ronald Getty, J. Paul
Getty Jr. David Hecht is deceased, Gordon removed.
George is given two votes.

Codicil No. 5, 6 March 1963

All personal wearing apparel, jewellery, ornaments and
all other articles of personal use to J. Paul Getty Jr,
George F. Getty II and J. Ronald Getty equally.

All shares of common stock registered in the name of Art
Properties Inc. are given to the trustees of the J. Paul
Getty Museum.

Art objects not suitable for exhibition to the public to go
to J. Paul Getty Jr, George F. Getty II and J. Ronald
Getty.

Getty Oil stock: 10,000 shares to J. Ronald Getty, 20,000
shares to George F. Getty II, 20,000 shares to J. Paul

Getty Jr. Gordon Peter Getty to receive $500 and nothing else.

Codicil No. 6, 16 September 1965

	Shares	Monthly allotment
Hildegarde Kuhn		$100
Penelope Kitson	5000	$1167
Mary Teissier	2000	$750
Robina Lund	5000	$750
Gloria Bigelow	500	$400
Mary Maginnes	500	$400
Belene Clifford	500	$300
Karin Mannhardt	200	$200

Executors remain the same.

Codicil No. 7, 11 March 1966

	Shares
Penelope Kitson	9500
Robina Lund	5000

Executors remain the same.

Codicil No. 8, 5 January 1967

Executors granted permission to abandon Italian villa unless they wish to occupy it. Leaves Italian villa to three sons (George F. Getty II, Ronald Getty, J. Paul Getty Jr) and arranges $500,000 to manage it.

Codicil No. 9, 3 November 1967

	Shares	Monthly allotment
Penelope Kitson	7500	$1167
Robina Lund	6250	$750
Marie Teissier	2500	$750
Gloria Bigelow	625	$400

Mary Maginnes	625	$400
Belene Clifford	625	$300
Karin Mannhardt	250	$200
Hildegarde Kuhn		$100
George F. Getty II	2000	
J. Ronald Getty	2000	
J. Paul Getty Jr	2000	
Gordon Getty		£500

Executors remain the same; George F. Getty II retains two votes.

Codicil No. 10, 24 February 1969

	Shares	Monthly allotment
Penelope Kitson	1000	$209
Robina Lund	1000	$209

Codicil No. 11, 28 March 1969

	Shares	Monthly allotment
Penelope Kitson	7500	$1167

Codicil No. 12, 26 June 1970

	Shares	Monthly allotment
Penelope Kitson	7500	$1167
Marie Teissier	2500	$750
Gloria Bigelow	625	$400
Mary Maginnes	625	$400
Belene Clifford	625	$300
Karin Mannhardt	250	$200
Robina Lund	1000	$209
Hildegard Kuhn		$100

Codicil No. 13, 8 March 1971

	Monthly allotment
Marianne von Alvensleben	$1000

Sutton Place employees	*Months' wages*
Charles Lee (chauffeur)	six
Frances Bullimore (butler)	six
Frank Parks (footman)	six
Katharine Aepli (cook)	six
Barbara M. Collings (secretary)	six
Albert Thurgood (estate manager)	six
Carole M. Tier (secretary)	six
Huchina Bannerman	six
Albert Sherman	six
Ronald H. Dean	three
Ernest E. Grafton	three
Sidney J. Fuller	three
Ronald E. Legg	three
Gertrude R. Munson	three
Norris Bramlett (administrative assistant)	six

Codicil No. 14, 29 July 1971

Executors: George F. Getty II, J. Ronald Getty. J. Paul Getty Jr (a.k.a. Eugene) removed. George and Ronald receive 2000 shares each. Gordon and J. Paul Jr receive $500 each. Security Pacific National Bank is appointed additional trustee. George Getty still has two votes.

Codicil No. 15, 20 March 1973

Penelope Kitson   5000 shares

Codicil No. 16, 14 June 1973

Executors: J. Ronald Getty and the Bank of America. George F. Getty II deceased 6 June 1973.

Codicil No. 17, 9 October 1973
Lady Ursula d'Abo   1000 shares

Codicil No. 18, 4 July 1974

Increase Sutton Place employee bequests to one full
   year's wages each.

Codicil No. 19, 21 January 1975

Executors: J. Ronald Getty, Gordon Getty and Title
   Insurance and Trust Co.

Codicil No. 20, 27 August 1975

	Shares	Monthly allotment
Penelope Kitson	5000	$1167
Marie Teissier	2500	$750
Gloria Bigelow	625	$400
Mary Maginnes	625	$400
Belene Clifford	625	$300
Karin Mannhardt	250	$200
Ursula d'Abo	1000	
Rosabella Burch	500	
Hildegard Kuhn		$100
Robina Lund		$209
Marianne von Alvensleben		$1000

Codicil No. 21, 11 March 1976

Entire estate bequeathed to trustees of J. Paul Getty
   Museum.

Appendix IV
Notes and Sources

For this biography of J. Paul Getty I have used material from personal reminiscences gathered in well over a hundred interviews with friends, acquaintances and members of the Getty family. I have also quoted J. Paul Getty himself as reported in his various autobiographies, articles and books. Other material is taken from the records of evidence submitted in the many court cases in which Getty and members of his family were involved over the years. There are a few instances where I have used information of a personal nature without identifying the speaker to protect my informants' anonymity. Otherwise I have tried to report Getty's life through his own voice and the voices of those who knew him well.

CHAPTER 1
Based on interviews with the Duke of Bedford, Jeannette Constable-Maxwell, Lady Diana Cooper, Gail Getty, J. Paul Getty Jr, Anna Hladka, Penelope Kitson, Barbara Wallace, Paul Louis Weiller; the Duke of Bedford's Memorial Address at the American Church of St Mark, North Audley Street, London, 21 June 1976; J. Paul Getty's Last Will and Testament, including twenty-one codicils, 1958–76, Los Angeles Superior Court, Probate Section, no. 622826.

The disappointment experienced by certain people over Getty's will is summed up in a letter written by one Getty aide to another: 'There were many people like you, people to whom Mr Getty made promises or clearly led them to believe that he would remember them in his will, but then made no provision. He imposed on so many people, allowed them to perform services for him with hardly even a thank you from him. There is only one word that describes Mr Getty and says it all – greed.'

Leon Turrou, Getty's consultant on security matters, claims that Getty promised several times to leave him $1 million in his will. When he arrived at Sutton Place from Paris on 6 June 1976 and found that he had got nothing, Turrou could sense 'Paul gloating at me, that he had pulled off one last ruse.'

CHAPTER 2

Robert B. Carney, *Hennepin County History: The Boyhood Years of J. Paul Getty in Minneapolis*, Summer, 1983.

Barbara Flanagan, *Minneapolis*, Nodin Press, 1973.

Jean Paul Getty, *As I See It*, Prentice Hall, 1976.

J. Paul Getty, *My Life and Fortunes*, Duell, Sloan & Pierce, 1963.

David Lavender, *California, A Bicentennial History*, Norton, 1976.

William Rintoul, *Spudding In*, California Historical Society, 1976.

Other sources include J. Paul Getty's birth certificate and the affidavit to amend the birth record, 7 January 1942.

On the choice of Christian names it was said in the Memorial Statement for George F. Getty that Paul is a 'stalwart, scriptural name'.

On George Franklin Getty's address at the time of his son's birth, Robert B. Carney was told by a Dr Eitel in Minneapolis that his father, also a doctor, had stayed at the West Hotel and 'often had meals with George Getty at the West End restaurant' in 1893 (Robert B. Carney, correspondence with the author, 26 April 1982, 27 June,

17 September 1983). This supports the possibility that Sarah and the infant J. Paul were living elsewhere. The Gettys lived at twelve different addresses in Minneapolis between 1885 and 1906.

For George Getty's studies at Ohio Northern, see Sarah Lehr Kennedy, *H. S. Lehr and His School*, which contains a letter from George Getty to Mrs Henry Soloman Lehr, the daughter of Ohio Northern's founder.

Information on George Getty's trip into Osage County with Frank Finney comes from Robert B. Finney, director, Public Affairs, Phillips Petroleum Company, letter to author, 12 November 1982.

For Henty, see Guy Arnold, *Held Fast for England: G. A. Henty, Imperialist Boys' Writer*, Hamish Hamilton, 1980. Many of Henty's tales were about Victorian empire-building in India, Africa and the Far East. None of the adolescent boy heroes have any love interest, and the fathers of Henty's heroes are usually 'curious recluse characters of weak or erratic dispositions who give a virtually free hand to their sons to go their own ways or organize and save the family fortunes.'

Details of Getty's academic career come from University of California, Berkeley, Office of Admissions and Records, Academic Record, and author's correspondence with the University of Oxford, 5 September, 1 November 1983.

For Getty's relations with his father, see correspondence between George F. Getty and J. Paul Getty, 1912–14. Much of their communication is taken up with money matters, such as the cost of operating a car on the Continent. Getty displayed an uncanny ability to analyse the smallest financial problem and argued his case trenchantly.

CHAPTER 3

American Association of Petroleum Geologists, *Trek of the Oil Finders: A History of Exploration for Petroleum*,

George Banta Co. Inc., Tulsa, Oklahoma, 1975. The Gettys are mentioned only seven times in the entire volume whereas Harry Sinclair, another early oilman, is mentioned fifty-six times.

George F. Getty, correspondence with the Bureau of Indian Affairs, Department of the Interior, 1911–19.

J. Paul Getty, *As I See It* and *My Life and Fortunes*.

J. Paul Getty, *The History of the Oil Business of George F. Getty and J. Paul Getty, 1903–1939*, privately published, 1941. Every single transaction in the first four decades of the Getty oil enterprises is here laboriously recorded. Getty obviously thought the book was an important document because it was printed on parchment paper bought from the Vatican and bound in ramskin supposed to last for 1000 years. Getty's literary effort was a bid for immortality.

Bob Gregory, 'The Legendary Men of Oil', *Oklahoma Monthly*.

Ruth Sheldon Knowles, *The Greatest Gamblers: The Epic of American Oil Exploration*.

Kenny A. Ranks and Paul F. Lambert, *Early Oklahoma Oil: A Photographic History*, 1981.

For Elsie Eckstrom, see *Elsie Eckstrom* vs *Paul Getty*, Los Angeles Superior Court, complaint filed 24 June 1917; the birth certificate for Elsie Eckstrom's daughter, unnamed, California State Board of Health, 1 December 1917; *Los Angeles Times*, 6 December 1917, 7 February 1918.

CHAPTER 4

Interviews with Teddy Hayes, Mark Kennedy, Ann Light (*née* Rork), Garth Young.

P.H. Frankel, *Essentials of Petroleum*, London, 1945.
J. Paul Getty, *As I See It, My Life and Fortunes* and *The History of the Getty Oil Business*.
Gene Tompkins, *California Adventures in Oil*, Aegean Park Press, 1981.

For Getty's relations with his first wife, see *Jeannette Getty* vs *J. Paul Getty*, Los Angeles Superior Court, petition no. D39912 filed 19 September 1925.

On Sarah Getty's fears about her son's godlessness, Mary S. Kennedy, granddaughter of the founder of Ohio Northern, told the author in July 1983 that Getty 'got clean away from God. He was godless. He cared about nobody but himself. He sold out to the devil.' Getty would repeat his father's homilies like 'moral responsibility must on no account be sidetracked,' but then be unable to follow them through.

Charles Wrightsman described Getty as a 'suspicious and secretive' man (letter to E. L. DeGolyer, 22 December 1928, DeGolyer Library, Southern Methodist University, Dallas, Texas).

Of her first meeting with Getty, Hildegard Kuhn told a German newspaper that in 1930 'an elegant man came up to me and asked me for a dance.' She refused, but he found out who she was and pursued her. Over the years he took her on several trips around Europe and sent her food parcels during the war. She never married and said, 'Perhaps he was partly to blame.' (Quoted in the *New York Times*, 15 June 1976.)

In the Memorial Statement for George F. Getty, he was said to be 'a plain man; unobtrusive, no show. His simplicity was akin to sublimity. . . . No pretence, no pretext, no duplicity.' His business philosophy 'was to help his fellow men. With Mr Getty the making of money was incidental.'

CHAPTER 5
J. Paul Getty, *As I See It* and *My Life and Fortunes*.
J. Paul Getty, *How to Be Rich*, Playboy Press, 1965. (Based on a series of articles which appeared in *Playboy*, 1961–5.)
James Grant, *Bernard Baruch*, Simon & Schuster, 1983.
Ralph Hewins, *The Richest American*, Dutton, 1960.

In *The History of the Getty Oil Business* Getty describes how he pulled off a coup to convince the Internal Revenue Service that his father's estate was not worth $20 million, as they claimed, but only $10 million. Again, the economic slump was the crucial factor in reducing the value of his father's holdings and thus the estate taxes. From that point on Getty arranged the family's business affairs so as to pay the least amount of estate tax possible.

In an unsigned letter to Frederick Wirth Jr in Berlin dated 19 March 1932, Getty's secretary wrote that Getty's income in 1931 was only $19,524.32, in addition to a capital gain of $12,478.98. This was inaccurate because Getty's tax return shows that income of $56,822.03 was reported to the Revenue Service.

On Getty's attempt to expand his oil interests in the Middle East, see 'Getty Oil Co.', National Archives, Diplomatic Branch, file 890G–6363.

CHAPTER 6
Interviews with Gordon Getty, Ann Light (*née* Rork), David Staples.

J. Paul Getty, *As I See It* and *My Life and Fortunes*.
Ralph Hewins, *The Richest American*.
W. A. Swanberg, *Citizen Hearst*, Charles Scribner's Sons, 1961.

On Getty's treatment of Ann, see *Helen Ann Rork Getty* vs *Jean Paul Getty*, Los Angeles Superior Court, petition no. D132457 filed 27 May 1935; *Helen Ann Rork Getty* vs *Jean Paul Getty*, Reno, Nevada, March 1936; *Los Angeles Times*, April–December 1935, March 1936. Paul Jr's birth was registered at the American Consulate, Passport Division, Genoa, Italy, on 9 September 1932.

For Ronald Getty's testimony about his parents' divorce, see *Jean Ronald Getty* vs *Gordon Peter Getty*, Los Angeles Superior Court, no. C286401, January 1984.

In 1933 Getty became so incensed with Otto Helmle's lawyer that he wrote to his German lawyer: 'Probably friends of mine, in Berlin, can take the matter up with the Hitler people, as I believe that Hitler is an honest man – very decided in his ways and wishes to see justice done without any delay.'

CHAPTER 7
Interviews with Gordon Getty, Howard Jarvis, Moses Lasky, Ann Light (*née* Rork).

On the establishment of the trust, see 'Declaration of Trust, Sarah C. Getty and J. Paul Getty', 31 December 1934; also, *Gordon Peter Getty* vs *J. Paul Getty*, San Francisco Superior Court, no. 570527, 1966, decided 17 April 1977; *Gordon Peter Getty* vs *J. Paul Getty*, California Court of Appeal, no. 28CA 996, decided 1972.

CHAPTER 8
J. Paul Getty, *As I See It, My Life and Fortunes, The History of the Getty Oil Business*.

On the battle for Tide Water, see 'Agreement of Merger between Tide Water Associated Oil Company, Tide Water Oil Company and Associated Oil Company', 15

October 1936; *Pacific Western Oil Corporation* vs *Mission Corporation*, Second Judicial Court, Reno, Nevada, no. 55691, 25 March 1937; *Los Angeles Times*, 30 April, 8 May 1936, 26 March 1937.

The Temporary National Economic Committee studies on industrial concentration in the 1930s made little of Getty. In 1937 he controlled a large block of Tide Water Associated, the ninety-second largest corporation in the USA (see Ferdinand Lundberg, *The Rich and the Super-Rich*, Bantam Books, 1968).

CHAPTER 9
Interviews with Joe Cumberland, Teddy Gaston (*née* Lynch), Ann Light (*née* Rork), Ware Lynch.

Eleanor Roosevelt, *This I Remember*, Harper, 1949.
William L. Shirer, *Berlin Diary*, Penguin Books, 1940.
William L. Shirer, *The Rise and Fall of the Third Reich*, Simon & Schuster, 1959.
'The Schiff Collection', *Art News*, 4 June 1938.
'The Art News of London', *Art News*, 14 July 1938.
Connoisseur, August 1938.
'View of the Year's Sales', *Connoisseur*, October 1938.

The June Hamilton Rhodes file in the Franklin Delano Roosevelt Library, Hyde Park, New York, includes long, detailed letters to Eleanor Roosevelt who was matron of honour at Rhodes's second wedding in 1933. (See also the William Gaston file and the J. Paul Getty file in the Franklin Delano Roosevelt Library.)

After Hitler's speech at the Sportspalast in late September 1938 Getty wrote in his diary: 'Führer makes great, manly speech. Crowd greets him uproariously.' Later, for publication the wording was changed to 'a speech in favor of peace'. (Entries for 27 August–14 November 1938.)

In an interview with the author Hilde Kruger claimed that Getty would come to visit her at the film studios and send her flowers. She had lunch with Hitler at the Berchtesgaden and thought him 'charming'. Later, in America, she claimed that Getty 'was afraid of rejection and told me he could never pay attention to me.' Kruger found him to be 'more interested in archaeology and sculpture than people.'

Getty's diary entry for the day he married Teddy says: 'Married at noon in a romantic setting, a palatial room in the Campidoglio. Lunch at the Ambassador with T. & J. Spent afternoon packing. Dinner with T. Leave for Naples at 8:30. T. & J. come to station to see me off. Much disappointed that T. decided not to go. Arr. Naples at 11:30 drive to Excelsior, fair room on bay for 60 l[ira]. Take short walk along Via Partenope & to bed.' (See entries for 24 March–30 December 1939.)

CHAPTER 10
Interview with Teddy Hayes by Doug Dodd, 4 December 1974. James Forrester correspondence, Princeton University.

Charles Higham, *Trading with the Enemy*, Delacorte Press, 1983.

Betty Kirk, *Covering the Mexican Front*, University of Oklahoma Press, 1942.

Walter J. Levy, *Oil Strategy and Politics, 1941–1981*, Westview Press, 1982.

For the FBI investigations of Getty and Hilde Kruger, see 'Jean Paul Getty', Federal Bureau of Investigation, confidential file no. 100–1202, 1940–73; 'Hilde Kruger',

Federal Bureau of Investigation, file no. 65–1157, National Archives, Diplomatic Branch; 'Kruger, Hilde', file no. 862.20211, National Archives, Diplomatic Branch.

Charles Higham was instrumental in alerting me to the FBI investigation of Getty and Hilde Kruger. I appreciate his kind advice and suggestions. Getty did not have commercial attachments with Nazi Germany to the extent of Standard Oil of New Jersey or Texaco.

For the FBI investigation of Charles Chaplin, see 'Charles Spencer Chaplin', Federal Bureau of Investigation, file no. 31–4467, 1943. Getty, interviewed by the FBI, admitted giving Joan Barry $150 a month 'until she got a job in pictures'. He declared: 'I have tried to help about a hundred people in the last 25 years. . . . Always made them sign a receipt saying they have no claim against me, my mother or wife.' Denying he slept with Barry, Getty told the FBI: 'I don't claim to be a saint but I am skittish about being bold. The less a man has on his conscience the better off he is.'

For the lawsuit against his mother, see *J. Paul Getty* vs *Sarah C. Getty and others*, Los Angeles Superior Court, no. 452368, 31 May 1940.

Sarah Getty's Last Will and Testament was filed at the Los Angeles Superior Court on 12 January 1942.

For Teddy's arrest in Italy, see 'Louise Lynch Getty interred in Italy', file no. 811.91265/142, National Archives, Diplomatic Branch.

For the confidential memorandum prepared by Naval Intelligence, see Office of Naval Intelligence to Fletcher Warren, Department of State, 22 March 1942.

At this point in his life Getty was 'buoyantly Chiliastic'. However, he later came to be in favour of enforced farmwork for 'Welfare parasites' and enforced abortions for women who became pregnant 'without official sanction'. In *As I See It* he predicted that the USA would pass legislation to ensure zero population growth.

CHAPTER 11
Interviews with Walter Bishop, Teddy Gaston (*née* Lynch), C. Lansing Hays Jr, Hilde Kruger, Ware Lynch, Maxie Sickinger, David Staples, Harold Stuart.

Charles Chaplin, *My Autobiography*, Simon & Schuster, 1964.
J. Paul Getty, *Europe in the Eighteenth Century*, privately published, 1947.
Ralph Hewins, *The Richest American*.
Roberta Louis Ironside, *An Adventure Called Skelly*, Appleton-Century-Crofts, 1970.

For the interview with the jury members at Chaplin's trial, see *Los Angeles Times*, 14 April 1945.

On his facelifts, Getty reports in his diary (10 May 1939) that his first facelift, costing $250, made him look five years younger than his age of forty-seven. The doctor's wife thought he looked fifteen years younger.

Getty was lucky to remain an oilman. In *Europe in the Eighteenth Century* he concludes: 'Taking it all in all, life in the 18th century was both harder and easier than life today . . . All classes had a serene faith in God and a cheerful outlook on life that should be an inspiration to their descendants.' He could walk down a Paris street and imagine aloud what great historical events took place there.

For Skelly's opposition to Getty, see *William G. Skelly vs Mission Corporation*, Reno, Nevada, docket 669, 4 November 1947. Information was also supplied by George H. Whitney, 21 December 1983.

On the trusteeship question, Getty wrote to his son George in January 1948 asking him 'in all humility' to step down as trustee of the Sarah Getty Trust and let his father return. Getty's plea for 'one competent, experienced and honest oilman' as trustee is a piece of classic dissembling, because he really needed to be the sole trustee to sell the trust's shares without George's opposition.

CHAPTER 12

Interviews with Joe Cumberland, Don Carlos Dunaway, Gordon Getty, John Pochna, Everette Skarda, Jack Sunderland, Harrison Symmes, Paul Walton, Garth Young.

Robert Lacey, *The Kingdom*, Hutchinson, 1981.

Stephen J. Longrigg, *Oil in the Middle East*, Oxford University Press, 1955.

John F. Mason, 'Petroleum Developments in Middle East Countries in 1965', *Bulletin of the American Association of Petroleum Geologists*, August 1966.

Leonard Mosley, *Power Play: Oil in the Middle East*, Random House, 1973.

Anthony Sampson, *The Seven Sisters*, Viking Press, 1975.

Secret History of the Oil Companies, introduced by William J. Kennedy, Documentary Publications, 1979.

Walter A. Tompkins, *Little Giant of Signal Hill: An Adventure in American Enterprise*, Prentice Hall, 1964.

In *The Seven Sisters* Anthony Sampson concludes that Getty was 'the first individual to challenge the dominance of the seven sisters in the Middle East, thus helping to begin the erosion of their monopoly.'

On the Neutral zone concessions, see memoranda from US officials in Saudi Arabia and Kuwait about Pacific Western Oil and Aminoil, Department of State, file 886A.2553/6–1653; also Pacific Western annual reports, 1948–56, and Mission Development annual reports, 1948 onwards.

CHAPTER 13
Interviews with Marianne von Alvensleben, the Duchess of Argyll, the Duchess of Bedford, Art Buchwald, Teddy Gaston (*née* Lynch), Gail Getty, J. Paul Getty Jr, Ware Lynch, John Pochna, Mary Teissier, Paul Louis Weiller, Frederico Zeri.

Bernard Berenson, *Sunset and Twilight*, Harcourt, Brace & World, 1983.
Mark Goulden, *Mark My Words!*, W. H. Allen, 1978.
Ethel LeVane and J. Paul Getty, *Collector's Choice*, W. H. Allen, 1955.

Getty's attraction for women was persistent. In 1937 a woman admirer sent him a poem and recalled 'the full, glorious memory of the years of friendship, a friendship which never loses one iota of its beauty, its luster, its strength.'

A letter from Getty's painter friend in 1949 reveals the freewheeling sexual life Getty was enjoying at the time. 'The two girls I asked on your behalf have been waiting . . . One of them is related to the Royal House and the other *their* great friend.'

According to Cleveland Amory (*Who Killed Society?*, Harper, 1960), Getty was a fringe member of international café society; he never belonged to the best circles. Amory visited Getty once in the George V Hotel and found him to be an unkempt, banal figure who 'looked like someone's aunt'.

Getty's attitude to money is partly explained by his diary entry for 27 February 1951: 'Lunch with Lloyd Gilmour (financier). He said I must be one of the half dozen richest men in the world? I said, yes, in stockholdings and underground oil reserves, but not in cash. He said, what difference does it make – except for the cash collector? We both laughed.'

Information on the Simons comes from an interview with Mrs Simon by Alexandra Bennion.

For Getty's accommodation at the George V, see *Time*, 24 February 1958.

Diana Cooper described Paul Louis Weiller as 'the frog who people can't endure', making him an outsider, like Getty. She met Getty through Paul Louis and persuaded him to give her a new door for her home as a present (Philip Ziegler, *Diana Cooper*, Hamish Hamilton, 1981).

For Teddy's divorce from Getty, see *Louise Getty* vs *Jean Getty*, Los Angeles Superior Court, final judgement of divorce, no. SMD13941, 28 May 1958.

For Timmy's death, see *Los Angeles Times*, 21 and 23 August 1958.

CHAPTER 14
Interviews with Art Buchwald, Joe Cumberland, Stuart Evey, Everette Skarda, Mary Teissier, Paul Louis Weiller, Garth Young.

On Tidewater expansion, see Tidewater Associated Oil Company prospectus to borrow $50 million, 26 March 1958.

Getty's oil production in the Neutral Zone was a very small factor in overall Middle Eastern oil production. Even the Japanese operator in the Neutral Zone, Arabian Oil Company, produced more barrels of oil than Getty Oil Co. However, in 1958, when the Neutral Zone reserves were estimated at over 12 billion barrels (50 per cent belonged to Getty), ARAMCO had oil reserves of 35 billion barrels, or a sixth of the world's total oil reserve.

See also Tidewater, Pacific Western and Getty Oil annual reports from 1951 to 1957.

CHAPTER 15
Interviews with the Duchess of Argyll, the Duke and Duchess of Bedford, Claus von Bulow, Jeannette Constable-Maxwell, John and Annabelle Pochna, Mary Teissier, Leon Turrou, Barbara Wallace, Paul Louis Weiller.

Art Buchwald, 'Mr Getty Watches the Tab', *Herald Tribune*, 24 November 1959.
J. Paul Getty, *As I See It* and *My Life and Fortunes*.
'Incredible Billionaire,' *Newsweek*, 7 March 1960.
Robina Lund, *The Getty I Knew*, Sheed, Andrews & McMeel Inc., Kansas City, 1977.
'Quite Social, Eh?', *Newsweek*, 11 July 1960.
Jacqueline Thompson, *The Very Rich Book*, Quill, 1981.
Time, 26 October 1959.

According to Robina Lund, Getty 'was such an appalling psychologist that he was quite unable to distinguish between the friends who did use him and those who wanted to give, not take.' He was 'so unimaginative by nature that he could not conceive of a situation that he himself had not experienced' (*The Getty I Knew*).

CHAPTER 16
Interviews with Norris Bramlett, Claus von Bulow, Gordon and Ann Getty, John Houchin, Moses Lasky, William Newsom, Everette Skarda, David Staples, Jack Sunderland, Leon Turrou.

John M. Blair, *The Control of Oil*, Pantheon, 1976.
Energy Policy in Perspective, Brookings Institute, 1981.

J. Paul Getty, *As I See It*.

J. Paul Getty, *How to Be a Successful Executive*, Playboy Press, 1971.

John McDonald, 'Paul Getty's Changed Plans', *Fortune*, December 1967.

William Wright, *The von Bulow Affair*, Delacorte, 1983.

Interviews with Sidney Petersen and Stuart Evey support the notion that Getty was not an astute manager. Petersen says that Getty 'had little knowledge or appreciation of how large and complicated the company had become.' Moreover, he was to blame for a sense of 'bureaucratic inaction'. He 'failed to realize the flood of foreign imports would depress the industry and make the tanker fleet almost worthless.' Evey says that the Pierre Marques Hotel near Acapulco was losing money in the 1960s because Getty, so many thousands of miles away in Britain, paid no attention to the operation. 'It was part of Getty Oil that no one was riding herd on.' The operation was improved by Evey, working for George II, and ultimately the hotel was sold to Daniel Ludwig at a profit.

Information from Claus von Bulow is contained in a memorandum to the author, 18 October 1984.

On the imposition of a quota system see memorandum from Gerald D. Morgan, 3 March 1958, and minutes of Cabinet meetings, 21 March 1958 and 6 March 1959, Eisenhower Library, Abilene, Kansas. Admiral Lewis Strauss stated that 'noncompliance by a few companies has tended to demoralize those that did comply.'

Getty also contributed to Eisenhower's 1952 election campaign. A telegram from Eisenhower to Getty in July 1952 reads: 'Your son delivered to me your very generous contribution. This is support which inspires me to carry forward in this fight for the principles in which we both believe.'

J. Ronald Getty's consulting agreement with Getty Oil Co., 20 November 1970, mentions the French lawsuit, *Société Nouvelle des Huiles (SNHM)* vs *Messrs Jouannic, Lemoine, Hebert and J. Ronald Getty*. Getty Oil Co. paid Ronald $450 a day to attend the trial since he was no longer a company employee.

Copies of the correspondence between Getty, Gordon and George II were provided by Getty Oil Co. The reference to 'blackmail' by Gordon was in George's handwritten notes, apparently made during a telephone conversation with his father. The Getty system of denial was operating during this interfamily donnybrook. Getty insisted: 'I do not try to bring business into my family life,' obviously an impossible task. Gordon, for his part, said later: 'When he claimed to be angry with me, I knew damn well he wasn't.' This overlooks the fact that he was nearly disinherited. His father wrote to him, quoting Gordon's mother, that 'at twenty-eight Gordon has retained the perspectives of an eight-year-old, i.e. that the nursery is the world and the sun beams for the benefit of the occupants.'

Some thirty-five years after buying the first Tide Water shares, Getty merged the larger company, which had $696 million in sales, into Getty Oil, which had a revenue of only $78 million. Getty felt that he created a stronger oil company, with one set of officers and directors instead of two separate ones. Yet, true to form, he was inconsistent, since he left Skelly Oil as a separate company, owned 71 per cent by Mission Corporation, which was in turn 65 per cent owned by Getty Oil.

In a letter to Thomas Dockweiler dated 5 July 1940, Hecht recommends that article 5 of the trust be amended so that 'stock dividends regularly declared and paid in lieu of cash dividends shall be income.' It appears that Getty may have wanted to obtain the stock dividends himself at that point. Perhaps this explains the reason for the amendment in 1941 favouring the income beneficiaries over the remaindermen.

In *How to Be a Successful Executive*, Getty estimated that by 1971 there were 120,000 millionaires in the USA, 105,000 more than in 1948.

CHAPTER 17
Interviews with the Duchess of Argyll, Harold Berg, Thomas Biamonte, Claus von Bulow, Jeannette Constable-Maxwell, Stuart Evey, Gordon and Ann Getty, Gail Getty, J. Paul Getty Jr, Jacqueline Getty Phillips, William Newsom, Barbara Wallace.

Tony Sanchez, *Up and Down with the Rolling Stones: The Inside Story,* New American Library, 1979.

According to Tony Sanchez, after breaking up with Paul Jr, Talitha 'stopped using drugs, cut down on booze, started to eat proper meals again and took a lover – Johnny Braces.'

Jacqueline Getty Phillips has described George's black periods and his love of graveyards. He would visit her former husband's grave and say to him: 'I'm going to take good care of you [i.e. Michael Riordan].' George was highly romantic and used to write Jackie several love notes each day. When his favourite dog died in 1955, George, then thirty-one, wrote: 'Felix belonged to me; he was his Daddy's boy. And his Daddy loved him immensely.'

For details of George's death, see *Los Angeles Times*, June 1973; autopsy report 'George F. Getty II', no. 73–9419, 20 June 1973; press release from Thomas T. Noguchi, County of Los Angeles Chief Medical Examiner–Coroner, 27 August 1973; letter from Thomas

T. Noguchi, Chief Medical Examiner–Coroner, and Ralph M. Bailey, Chief, Investigations Division, to Paul Thomas Guinn, 9 November 1973.

For Getty, according to Giorgio Schanzer, after George's death 'something had broken. It represented his continuity, his survival. If George had survived, perhaps he could have created a great worldwide oil company.'

For details of George's will, see Estate of George F. Getty II, Los Angeles Superior Court, no. 594490, filed 15 June 1973.

For details of the kidnapping, see report of the Procura della Republica di Lagonegro, 1974, and *Il Messagero*, July–December 1973.

According to Paul Louis Weiller, during the kidnapping Getty told him that 'the family was ready to pay, but that he wouldn't pay a cent.' After the boy was ransomed Getty told Weiller that 'the family paid, but he wouldn't ever say that it was he who had done it.'

CHAPTER 18

Interviews with Sir Geoffrey Agnew, John Brealey, Burton Fredericksen, Jiri Frel, Stephen Garrett, Rudolf Heineman, Anna Hladka, Claude Sere, Baron H. H. Thyssen-Bornemisza, Francis Watson, Gillian Wilson, Frederico Zeri, Martin Zimet.

'British Nearing Goal of Keeping Titian Sold to Getty', *New York Times*, 4 June 1972.

Burton B. Fredericksen, *Masterpieces of Painting in the J. Paul Getty Museum*, J. Paul Getty Museum, 1980.

Burton B. Fredericksen, 'New Information on Raphael's *Madonna di Loretto*', *J. Paul Getty Museum Journal*, 1976.

David Gebhard, 'Getty's Museum', *Architecture Plus*, 1974.

J. Paul Getty, *The Golden Age*, Trident Press, 1968.

J. Paul Getty, *The Joys of Collecting*, Hawthorn Books, 1965.

'Getty Museum Gets Britain's Titian', *Herald Tribune*, 29 June 1971.

Cecil Gould, 'The Millionaire and the Madonna', *Connoisseur*, 1984.

Ed Hotaling, 'Herculaneum at Malibu', *Art News*, September 1971.

Interview with J. Paul Getty, *Art News*, September 1971.

J. Paul Getty Museum Journal, vol. 3, 1976.

David Shaw, 'J. Paul Getty's Dream Museum: Critics Pan It, Public Loves It', *Smithsonian*, May 1974.

'Welcome to Gettyland', *Newsweek*, 28 January 1974.

On *The Madonna of the Loretto*, Anna Hladka claimed, to the author on 26 March 1984, that she ordered a second photograph of the engraving from the British Museum in 1977. But she does not say if it was in fact sent. 'I remember clearly that I went to the Department of Prints and Drawings one day and the engraving was not in the folder, and it was a shock to me, but today I am not really sure [if] it was between 1972 and 1977 or between 1977 and 1978.' In any case, the engraving had somehow disappeared. Mrs Hladka says it was obvious from the engraving that Getty did not have the real *Madonna* because 'there is a difference apparent at first sight in the shape of the drapery and in the hairstyle of the Madonna.' Getty's 'Raphael' is no longer on display and is considered to be virtually worthless.

Mrs Hladka has described her relationship with Getty in intimate detail. One night, after she had escaped from Soviet-occupied Czechoslovakia, Getty asked her to come to his bedroom. Anna reflected, 'I am forty-five and free, and I like him.' She found Getty 'thinner and smaller than my husband' but 'he smelled sweet and his skin was soft.' She felt 'more like a mother than a lover,' yet Getty 'treated me like an equal, hugging me, talking to me.' His lovemaking was 'not simply a penetration, but friendly. He was always in control and often the lovemaking would last for two hours.' Getty, at the age of seventy-five, 'was at full physical strength and could always satisfy me.' Their lovemaking followed a regular pattern. After they had made love, 'he always went to the bathroom to wash his underwear and then hang it neatly to dry.' During this interlude Anna would make the bed. Then he would 'tell me a few political jokes, nothing in bad taste.' Together they might share a few chocolates. If there were flies buzzing about he might stand on the bed trying to swat them. Finally he would say, 'I think it is time we got some sleep.'

In the opinion of Sir Francis Watson, Getty could not compare as a collector with the Marquess of Hertford, the son of Sir Richard Wallace, who was responsible for putting together the Wallace Collection in London. Watson feels Hertford had 'the flair, the passion and an income equivalent to $600 million a year' to buy an art object every day.

CHAPTER 19
Interviews with the Duchess of Argyll, Harold Berg, Don Carlos, Gordon Getty, John Pochna, David Rockefeller, Everette Skarda.

John Blair, *The Control of Oil*, Random House, 1976.
Peter Collier and David Horowitz, *The Kennedys: An American Drama*, Summit Books, 1984.

Peter Collier and David Horowitz, *The Rockefellers: An American Dynasty*, Holt, Rinehart & Winston, 1976.

Robert Engler, *The Brotherhood of Oil*, University of Chicago Press, 1977.

Forget Not: The Autobiography of Margaret, the Duchess of Argyll, W. H. Allen, 1975.

J. Anthony Lukas, *Nightmare: The Underside of the Nixon Years*, Viking Press, 1976.

Goronwy Rees, *The Multimillionaires: Six Studies of Wealth*, Macmillan, 1961.

According to John M. Blair, by 1970 Getty Oil's ranking in the oil industry was ninth in ownership of crude oil reserves and production, sixteenth in refining capacity and below the top twenty in retail gasoline sales, where fierce price wars were being waged. However, internal Getty Oil documents show that the company was paying one of the lowest rates of tax in the industry, some 23.5 per cent on income (*The Control of Oil*).

For Getty's connection with Nixon, see Senate Select Committee, Watergate Hearings. According to Robert Engler, various oil sources contributed $5 million to Nixon, including illegal cash contributions from Gulf, Phillips and Ashland. Getty's friend, Armand Hammer, gave $100,000 (*The Brotherhood of Oil*).

According to Gordon Getty, 'Father called himself a capitalist but he meant entrepreneur. He was always a nonconformist, but you never could tell whether business was great or a catastrophe from the look on his face.'

CHAPTER 20

Interviews with Marianne von Alvensleben, the Duchess of Argyll, the Duke of Bedford, Harold Berg, Bela von Block, Norris Bramlett, Ann Cole, Jeannette Constable-Maxwell, Ann Getty, Gail Getty, Anna Hladka, Edward

Landry, Charles Lee, Dr Clive Mackenzie, Giorgio
Schanzer, Claude Sere, Mary Teissier, Barbara Wallace,
Frederico Zeri, Martin Zimet.

Mark Goulden, *Mark My Words!*
Robina Lund, *The Getty I Knew*.

According to Giorgio Schanzer, Getty was 'more a business machine who wasn't happy or unhappy. He was
singleminded about business. He was not a miser. He
just wasn't generous.'

John Brealey was sickened by the 'scheming and
intrigue that went on at Sutton Place.' He likened it to
the French Court of the eighteenth century, about which
La Bruyère had written: 'A king lacks nothing except the
sweetness of a private life.'

Getty refused to pay some women monthly allowances
unless they signed a release form each month. The form
read: 'I have absolutely no claim or possible claims,
whatever, of any kind, character or nature existing or
contingent against J. Paul Getty, and there are no matters
which exist to the date thereof which can arise or result
in such claims. Without implying the existence of any
claims, I hereby release J. Paul Getty from any and all
manner of claims of any kind or nature whatsoever which
I ever had, now have, or may hereafter have against
him.'

To help block the British Government from claiming
death duties, 'hardly a week went by without his return
to his home in Malibu being discussed. Not just as a
concept, but with concrete plans for the move' (Claus
von Bulow to Gordon Getty, 14 July 1977).

Even though he was fading fast, Getty on 30 April
bought a block of Mission Corporation shares; on 4 May
he approved the purchase of magnificent cabinets owned

by the Maharani of Baroda; and on 25 May he wrote in his diary that he was sorry that he was unable to attend the Chelsea Flower Show. He also noted that the Internal Revenue Service was sending a man to audit Sutton Place.

Paul Jr recalls, 'The only time Father spoke to me about mortality was to make sure Penelope would keep the cottage after he was gone.'

EPILOGUE

Interviews with Norris Bramlett, John Brealey, Claus von Bulow, Stuart Evey, Burton Fredericksen, Stephen Garrett, Ann and Gordon Getty, Gail Getty, J. Paul Getty Jr, Edward Landry, Moses Lasky, Martin Lipton, William Newsom, Sidney Petersen, Martin Siegel, Edward Stadum, Vanni Treves, Leon Turrou, John Walker, Barbara Wallace, Gillian Wilson.

Ivan Boesky, *Merger Mania*, Holt, Rinehart & Winston, 1985.

Patricia Failing, 'The $1,000,000 Art Museum', *Art News*, November 1981.

'The Golden Eye of Los Angeles', *Newsweek*, 26 November 1984.

Joseph Morgenstern, 'Getty's Little Place in Malibu', *Art News*, March 1977.

New York Times, 6 and 15 June 1976.

Roy Rowan, 'A Little Museum in Malibu Could Throw the Art Market for a Loop', *Fortune*, 3 July 1978.

Karl W. Schlubach, *Getty Oil Reports*, Hamershlag, Kempner & Co., 1973–83.

The Times (London), 7 June 1976.

Gillian Wilson, *Decorative Arts in the J. Paul Getty Museum*, J. Paul Getty Museum, 1977.

According to Gordon Getty, the last codicil was changed because Joe Peler, Getty's chief tax adviser in the USA, 'thought it . . . possibly dangerous.' Peler did not want the money to go to the trust because it had no power to do anything, and so the will was changed so that the money went directly to the museum.

Getty left money to three California women friends, even though he had not set foot in his home state for twenty-five years. Gloria Bigelow of Los Angeles and Mary Maginnes of Malibu each received $103,281 of stock and $400 a month for life. Getty also left $300 a month to Belene Clifford, the sister of his second wife, Allene, in addition to the same gift of stock.

In 'The Ins and Outs of J. Paul Getty's Baffling Bequests' by Bruce Page (*New West*, 20 December 1976), Robina Lund is quoted as saying: 'He would tell people they were in the will for so much, or they had been cut down, or were out altogether. I used to walk out on him when he ordered me about. . . . I told him I didn't want to be in his will at all if he was going to bring it up every time we had a disagreement.'

Getty always bragged that servants in Britain came cheap. In his last years he generously increased their bequests to a year's salary. But since Barbara Wallace was paid only $105 a week as his secretary, she received a mere $5471.45. Francis Bullimore, the butler, received $4135.40, and Gertrude Munson, the chief housekeeper, was left $1848.46.

For Ronald Getty's suit, see *Jean Ronald Getty et al.* vs *Gordon Peter Getty, etc. et al.,* no. C286401, 1981; transcript of trial, 9–22 January 1984.

For other lawsuits concerning the trust, see 'Declaration of Gail Harris Getty re Status of Seth Hufstedler as Guardian ad Litem of Tara Getty', Los Angeles Superior Court, case no. P685566, 18 April 1984; 'In the Matter of the Declaration of Trust of Sarah C. Getty', no. P685566, June 1984; affidavit of Cecil Eppel MD at US Embassy, 16 December 1983.

According to Jeannette Constable-Maxwell, Gordon is 'a font of natural wisdom, a storehouse of information. He would have made an excellent schoolteacher because of his enthusiasm for knowledge.' Sidney Petersen says, 'There was a problem in getting Mr Getty to focus on solutions for the company. He would always come up with other alternatives.' Petersen was afraid that Gordon wanted absolute control of the company, just as his father had.

A memorandum on the opening of the J. Paul Getty Museum in 1974 reveals that Getty himself originally came up with the idea to entice the 'Norton Simon Galleries' into some form of collaboration.

After Paul Jr's gift to the National Gallery was announced, the Prime Minister, Margaret Thatcher, wrote to him that 'the Cabinet agreed that the standing of the institution and the munificence of the gift combine to render this an occurrence of national significance and . . . express their warm appreciation of and profound gratitude for your magnificent generosity.' Mrs Thatcher subsequently visited Paul Jr in the London Clinic where they discussed his charitable donations. She also sent him a portrait of herself with a personal inscription.

On hearing of Paul Jr's gift to the National Gallery, Gordon wrote to his brother: 'Three cheers for your magnificent support. If that means Britain can keep art works that might have gone to JPGM (J. Paul Getty Museum] that's fine. If you personally were to target JPGM acquisitions for retention funds, picking on us and not on other purchasers, I don't see how I could find much fault with that either.'

In his interview with Barbara Walters on *20/20*, Gordon explained that his father was 'a first-rate, although Victorian father.' His own life he described as 'the same life with more zeros.'

Index

Index

I AM A SUCCESSFUL
AREA DISTRIBUTOR
WITH AN 100 +
ORGANISATION IN
JANURARY 1996.

— YOUR FREE ENTRY —

$500 CASH DRAW!

NEXT DRAW:
8th AUGUST 1995

Permit No. 95/1731

Have Your Carpet Spot & Deep
Cleaned or Shampooed.

In exchange for previewing an obligation-free presentation of
the NEW G4.

In addition, after the presentation is completed you will be
registered in our $500 Cash Draw. The draw will be published
in the Telegraph/Mirror on the 10th August, 1995.

Sponsored by: MALVALE PTY. LTD., 26 Marion Street, Parramatta. PHONE: 891 1322

Local Distributor: GILPORT PTY. LTD., 23 FERN ST., ISLINGTON **PH: (049) 61 3050**
ACN 003 495 181
Subject to company rules — Adults only — No obligation to purchase — strictly a marketing programme.
Husband and wife to both be present.

KEEP THIS CARD BY YOUR PHONE

The best in biography from Grafton Books

Henry Cecil On the Level (illustrated)	£2.50	☐
Bryan Robson United I Stand (illustrated)	£1.95	☐
Dudley Doust Ian Botham	£1.50	☐
Kitty Hart Return to Auschwitz (illustrated)	£2.50	☐
Roger Manvell and Heinrich Frankel Hitler: The Man and the Myth	£2.95	☐
Desmond Morris Animal Days	£1.95	☐
Axel Munthe The Story of San Michele (illustrated)	£3.95	☐
Professor Keith Simpson Forty Years of Murder (illustrated)	£2.95	☐
Elizabeth Longford The Queen Mother (illustrated)	£3.95	☐
Greville Wynne The Man from Odessa (illustrated)	£2.50	☐
Livia E Bitton Jackson Elli: Coming of Age in the Holocaust	£2.50	☐
Graham Gooch Out of the Wilderness (illustrated)	£2.50	☐

To order direct from the publisher just tick the titles you want
and fill in the order form.   GB581

All these books are available at your local bookshop or newsagent, or can be ordered direct from the publisher.

To order direct from the publishers just tick the titles you want and fill in the form below.

Name _____

Address _____

_____

Send to:
Grafton Cash Sales
PO Box 11, Falmouth, Cornwall TR10 9EN.

Please enclose remittance to the value of the cover price plus:

UK 60p for the first book, 25p for the second book plus 15p per copy for each additional book ordered to a maximum charge of £1.90.

BFPO 60p for the first book, 25p for the second book plus 15p per copy for the next 7 books, thereafter 9p per book.

Overseas including Eire £1.25 for the first book, 75p for second book and 28p for each additional book.

Grafton Books reserve the right to show new retail prices on covers, which may differ from those previously advertised in the text or elsewhere.